T0211181

Applied Mathematics with
Open-Source Software

Chapman & Hall/CRC Series in Operations Research
Series Editors:

Malgorzata Sterna, Bo Chen, Michel Gendreau, and Edmund Burke

For more information about this series please visit: https://www.routledge.com/
Chapman--HallCRC-Series-in-Operations-Research/book-series/CRCOPSRES

Applied Mathematics with Open-Source Software
Operational Research Problems with Python and R

Vincent Knight
Cardiff University, United Kingdom

Geraint Palmer
Cardiff University, United Kingdom

CRC Press
Taylor & Francis Group
Boca Raton London New York

CRC Press is an imprint of the
Taylor & Francis Group, an **informa** business

A CHAPMAN & HALL BOOK

First edition published 2022
by CRC Press
6000 Broken Sound Parkway NW, Suite 300, Boca Raton, FL 33487-2742

and by CRC Press
4 Park Square, Milton Park, Abingdon, Oxon, OX14 4RN

© 2022 Taylor & Francis Group, LLC

CRC Press is an imprint of Taylor & Francis Group, LLC

Library of Congress Cataloging-in-Publication Data

Names: Knight, Vincent (Vincent A.), author. | Palmer, Geraint, author.
Title: Applied mathematics with open-source software : operational research problems with Python and R / authored by Vincent Knight, Cardiff University, United Kingdom, Geraint Palmer, Cardiff University, United Kingdom.
Description: First edition. | Boca Raton : C&H/CRC Press, 2022. | Series: Chapman & Hall/CRC series in operations research | Includes bibliographical references and index.
Identifiers: LCCN 2021055603 (print) | LCCN 2021055604 (ebook) | ISBN 9780367348687 (hbk) | ISBN 9780367339982 (pbk) | ISBN 9780429328534 (ebk)
Subjects: LCSH: Operations research--Data processing. | Mathematics--Data processing. | Python (Computer program language) | R (Computer program language)
Classification: LCC T57.5 .K55 2022 (print) | LCC T57.5 (ebook) | DDC 658.4/0340285--dc23/eng/20220202
LC record available at https://lccn.loc.gov/2021055603
LC ebook record available at https://lccn.loc.gov/2021055604

ISBN: 978-0-367-34868-7 (hbk)
ISBN: 978-0-367-33998-2 (pbk)
ISBN: 978-0-429-32853-4 (ebk)

DOI: 10.1201/9780429328534

Typeset in Latin Modern font
by KnowledgeWorks Global Ltd.

Publisher's note: This book has been prepared from camera-ready copy provided by the authors.

Contents

SECTION V Optimisation

Authors

Vincent Knight
Cardiff University School of Mathematics
Cardiff, Wales, UK

Geraint Palmer
Cardiff University School of Mathematics
Cardiff, Wales, UK

I

Getting Started

CHAPTER 1

Introduction

T HANK you for starting to read this book. This book aims to bring together two fascinating topics:

- Problems that can be solved using mathematics;

- Software that is free to use and change.

What we mean by both of those things will become clear through reading this chapter and the rest of the book.

1.1 WHO IS THIS BOOK FOR?

This book is aimed at readers who want to use open-source software to solve the considered applied mathematical problems.

If you are a student of a mathematical discipline, a graduate student of a subject like operational research, a hobbyist who enjoys solving the travelling salesman problem or even if you get paid to do this stuff: this book is for you. We will introduce you to the world of open-source software that allows you to do all these things freely.

If you are a student learning to write code, a graduate student using databases for their research, an enthusiast who programs applications to help schedule weekly chores, or even if you get paid to write software: this book is for you. We will introduce you to a world of problems that can be solved using your skill set.

It would be helpful for the reader of this book to:

- Have access to a computer and be able to connect to the internet to be able to download the relevant software;

- Have done any introductory tutorial in the languages they plan to use;

- Be prepared to read some mathematics. The topics covered use some algebra, calculus and probability. Technically you do not need to understand the specific mathematics to be able to use the tools in this book.

By reading a particular chapter of the book, the reader will have:

DOI: 10.1201/9780429328534-1

1. the practical knowledge to solve problems using a computer;

2. an overview of the higher level theoretic concepts;

3. pointers to further reading to gain background understand and research undertaken using the concepts.

1.2 WHAT DO WE MEAN BY APPLIED MATHEMATICS?

We consider this book to be a book on applied mathematics. This is not, however, a universal term, for some applied mathematics is the study of mechanics and involves things like modelling projectiles being fired out of canons. We will use the term a bit more freely here and mean any type of real-world problem that can be tackled using mathematical tools. This is sometimes referred to as operational research, operations research, mathematical modelling or indeed just mathematics.

One of the authors, Vince, used mathematics to understand just how bad one of the so called "worst plays in Super Bowl history was". Using an area of mathematics called game theory (seen in Chapter 6), he showed that perhaps the strategic decision making was not as bad as it seemed, the outcome was just unlikely[1].

The other author, Geraint, used mathematics to find the best team of Pokémon. Using an area of mathematics called linear programming (seen in Chapter 8) which is based on linear algebra he was able to find the best makeup of Pokémon[2].

Here, applied mathematics is the type of mathematics that helps us answer questions that the real world asks.

1.3 WHAT IS OPEN-SOURCE SOFTWARE

Strictly speaking open-source software is software with source code that anyone can read, modify and improve. In practice this means that you do not need to pay to use it which is often one of the first attractions. This financial aspect can also be one of the reasons that someone will not use a particular piece of software due to a confusion between cost and value: if something is free, is it really going to be any good?

In practice open-source software is used all over the world and powers some of the most important infrastructure around. A good example of this is cryptographic software which should not rely on secrecy for security[3] This implies that cryptographic systems that do not require trust in a hidden system can exist. In practice these are all open-source.

Today, open-source software is a lot more than a licensing agreement: it is a community of practice. Bugs are fixed faster, research is implemented immediately and knowledge is spread more widely thanks to open-source software. Bugs are fixed

---

[1]At the time of writing this is available to read at: https://vknight.org/unpeudemath/gametheory/2015/02/15/on-the-worst-play-in-superbowl-history.html

[2]At the time of writing this is available to read at: http://www.geraintianpalmer.org.uk/2018/05/29/pokemon-team-pulp/

[3]This is also referred to as Kerckhoffs's principle which states that "a cryptosystem should be secure, even if everything about the system, except the key, is public knowledge" [32].

faster because anyone can read and inspect the source code. Most open-source software projects also have clear mechanisms for communicating with the developers and even reviewing and accepting code contributions from the general public. Research is implemented immediately because when new algorithms are discovered, they are often added directly to the software by the researchers who found them. This all contributes to the spread of knowledge: open-source software is the modern shoulder of giants that we all stand on.

Open-source software is software that, like scientific knowledge is not restricted in its use.

1.4   HOW TO GET THE MOST OUT OF THIS BOOK

The book itself is open-source. You can find the source files for this book online at `https://github.com/drvinceknight/ampwoss`. There you will also find a number of *Jupyter notebooks* and *R markdown files* that include code snippets that let you follow along.

We feel that you can choose to read the book from cover to cover, writing out the code examples as you go; or it could also be used as a reference text when faced with a particular problem and wanting to know where to start.

After this introductory chapter the book is split into 4 sections. Each section corresponds to a broad problem type and contains 2 chapters that correspond to 2 solution approaches. The first chapter in a section is based on exact methodology whereas the second chapter is based on heuristic methodology. The structure of the book is:

1. Probabilistic modelling

   - Markov chains
   - Discrete event simulation

2. Dynamical systems

   - Differential equations
   - Systems dynamics

3. Emergent behaviour

   - Game theory
   - Agent-based simulation

4. Optimisation

   - Linear programming
   - Heuristics

Every chapter has the following structure:

1. Introduction - a brief overview of a given problem type. Here we will describe the problem at hand in general terms.

2. An example problem. This will provide a tangible example problem that offers the reader some intuition for the rest of the discussion.

3. An overview of the theory as well as a discussion as to how the theory relates to the considered problem. Readers will also be presented with reference texts if they want to gain a more in-depth understanding.

4. Solving with Python. We will describe how to use tools available in Python to solve the problem.

5. Solving with R. We will describe how to use tools available in R to solve the problem.

6. The wider context. This section will include a few hand-picked academic papers relevant to the covered topic. It is hoped that these few papers can be a good starting point for someone wanting to not only use the methodology described but also understand the broader field.

For a given reader, not all sections of a chapter will be of interest. Perhaps a reader is only interested in R and finding out more about the research. The R and Python sections are **purposefully** written as near clones of each other so that a reader can read only the section that interests them. In places there are some minor differences in the text, and this is due to differences of implementation in the respective languages.

Note that the solution proposed to each problem in each chapter is not necessarily unique. For example, in Chapter 3 the Python library Ciw [52, 72] is used whereas an alternative could be to use a Python library called SimPy [73].

Please do take from the book what you find useful.

1.5 HOW CODE IS WRITTEN IN THIS BOOK

Throughout this book, there are going to be various pieces of code written. Code is a series of instructions that usually give some sort of output when submitted to a computer.

This book will show both the set of instructions (referred to as the input) and the output.

You will see Python input as follows:

```
Python input
1  print(2 + 2)
```

and you will see Python output as follows:

```
Python output
2   4
```

You will see R input as follows:

```
R input
3   print(2 + 2)
```

and you will see R output as follows:

```
R output
4   [1] 4
```

As well as this, a continuous line numbering across all code sections is used so that if the reader needs to refer to a given set of input or output this can be done. The code itself is written using 3 principles:

- Modularity: code is written as a series of smaller sections of code. These correspond to smaller, simpler, individual tasks (modules) that can be used together to carry out a particular larger task.

- Documentation: readable variable names as well as text describing the functionality of each module of code are used throughout. This ensures that code is not only usable but also understandable.

- Tests: there are places where each module of code is used independently to check the output. This can be thought of as a test of functionality which readers can use to check they are getting expected results.

These are best practice principles in research software development that ensure usable, reproducible and reliable code [84]. Interested readers might want to see Figure 1.1 which shows how the 3 principles interact with each other.

Thank you for picking up this book, we hope one if not all of the following chapters proves interesting or useful to you.

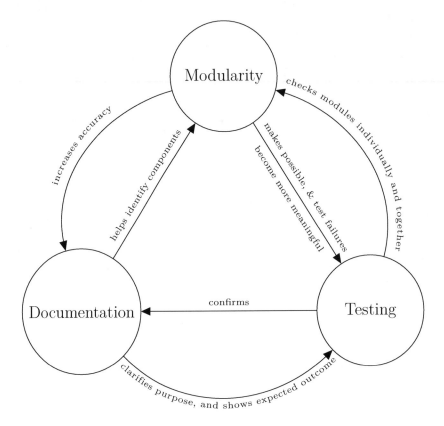

Figure 1.1 The relationship between modularisation, documentation and testing. The authors thank Dr Nikoleta E. Glynatsi for their contribution to the drawing of this diagram.

II

Probabilistic Modelling

CHAPTER 2

Markov Chains

M ANY real-world situations have some level of unpredictability through random-ness: the flip of a coin, the number of orders of coffee in a shop, the winning numbers of the lottery. However, mathematics can in fact let us make predictions about what can be expected to happen. One tool used to understand randomness is Markov chains, an area of mathematics sitting at the intersection of probability and linear algebra.

2.1   PROBLEM

Consider a barber shop. The shop owners have noticed that customers will not wait if there is no room in their waiting room and will choose to take their business elsewhere. The barber shop would like to make an investment so as to avoid this situation. They know the following information:

- They currently have 2 barber chairs (and 2 barbers);

- they have waiting room for 4 people;

- they usually have 10 customers arrive per hour;

- each barber takes about 15 minutes to serve a customer so they can serve 4 customers an hour.

This is represented diagrammatically in Figure 2.1.

They are planning on reconfiguring space to either have 2 extra waiting chairs or another barber's chair and barber.

The mathematical tool used here to model this situation is a Markov chain.

2.2   THEORY

A Markov chain is a model of a sequence of random events that is defined by a collection of **states** and rules that define how to move between these states.

For example, in the barber shop, a single number is sufficient to describe the status of the shop: the number of customers present. If that number is 1 this implies that

DOI: 10.1201/9780429328534-2

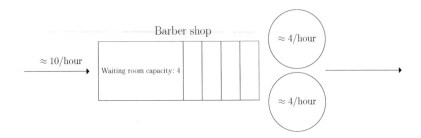

Figure 2.1 Diagrammatic representation of the barber shop as a queuing system.

1 customer is currently having their hair cut. If that number is 5, this implies that 2 customers are being served and 3 are waiting. The entire set of values that this value can take is a finite set of integers from 0 to 6, this set, in general, is called the *state space*. If the system is full (all barbers and waiting room occupied), then the Markov chain is in state 6 and if there is no one at the shop then it is in state 0. This is denoted mathematically as:

$$S = \{0, 1, 2, 3, 4, 5, 6\} \tag{2.1}$$

The state increases when people arrive and this happens at a rate of change of 10 per unit time. The state decreases when people are served and this happens at a rate of 4 per active server per unit time. In both cases it is assumed that no 2 events can occur at the same time.

In general, the rules that govern how to move between these states can be defined in 2 ways:

- Using probabilities of changing state (or not) in a well-defined time interval. This is called a discrete time Markov chain.

- Using rates of change from one state to another. This is called a continuous time Markov chain.

The barber shop will be considered as a continuous time Markov chain as shown in Figure 2.2

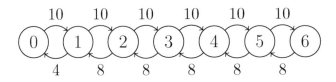

Figure 2.2 Diagrammatic representation of the state space and the transition rates.

Note that a Markov chain assumes the rates follow an exponential distribution. One interesting property of this distribution is that it is considered memoryless which means the probability of a customer finishing service within the next 5 minutes does not change if they have been having their hair cut for 3 minutes already.

These states and rates can be represented mathematically using a transition matrix Q where Q_{ij} represents the rate of going from state i to state j. In this case:

$$Q = \begin{pmatrix} -10 & 10 & 0 & 0 & 0 & 0 & 0 \\ 4 & -14 & 10 & 0 & 0 & 0 & 0 \\ 0 & 8 & -18 & 10 & 0 & 0 & 0 \\ 0 & 0 & 8 & -18 & 10 & 0 & 0 \\ 0 & 0 & 0 & 8 & -18 & 10 & 0 \\ 0 & 0 & 0 & 0 & 8 & -18 & 10 \\ 0 & 0 & 0 & 0 & 0 & 8 & -8 \end{pmatrix} \tag{2.2}$$

You will see that Q_{ii} are negative and ensure the rows of Q sum to 0. This gives the total rate of change leaving state i.

The matrix Q can be used to understand the probability of being in a given state after t time units. This can be represented mathematically using a matrix P_t where $(P_t)_{ij}$ is the probability of being in state j after t time units having started in state i. Using a mathematical tool called the matrix exponential,[1] the value of P_t can be calculated numerically.

$$P_t = e^{Qt} \tag{2.3}$$

What is also useful is understanding the long run behaviour of the system. This allows us to answer questions such as "what state is the system most likely to be in on average?" or "what is the probability of being in the last state on average?".

This long run probability distribution over the states can be represented using a vector π where π_i represents the probability of being in state i. This vector is in fact the solution to the following matrix equation:

$$\pi Q = 0 \tag{2.4}$$

with the following constraint:

$$\sum_{i=1}^{n} \pi_i = 1 \tag{2.5}$$

In the upcoming sections all of the above concepts will be demonstrated and used to understand what is the best course of action for the barber shop.

2.3 SOLVING WITH PYTHON

The first step is to write a function to obtain the transition rates between 2 given states:

---

[1]Chapter 9 of [77] gives a description of how to compute the matrix exponential numerically and [46, 47] give a review of 19 algorithms that can be used.

Python input

```python
def get_transition_rate(
    in_state,
    out_state,
    waiting_room=4,
    num_barbers=2,
):
    """Return the transition rate for 2 given states.

    Args:
        in_state: an integer
        out_state: an integer
        waiting_room: an integer (default: 4)
        num_barbers:  an integer (default: 2)

    Returns:
        A real.
    """
    arrival_rate = 10
    service_rate = 4
    capacity = waiting_room + num_barbers
    delta = out_state - in_state

    if delta == 1:
        return arrival_rate

    if delta == -1:
        return min(in_state, num_barbers) * service_rate

    return 0
```

Next, a function that creates an entire transition rate matrix Q for a given problem is written. The Numpy [28] library will be used to handle all the linear algebra and the Itertools library for some iterations:

Python input

```python
34  import itertools
35  import numpy as np
36
37
38  def get_transition_rate_matrix(waiting_room=4, num_barbers=2):
39      """Return the transition matrix Q.
40
41      Args:
42          waiting_room: an integer (default: 4)
43          num_barbers: an integer (default: 2)
44
45      Returns:
46          A matrix.
47      """
48      capacity = waiting_room + num_barbers
49      state_pairs = itertools.product(range(capacity + 1), repeat=2)
50      flat_transition_rates = [
51          get_transition_rate(
52              in_state=in_state,
53              out_state=out_state,
54              waiting_room=waiting_room,
55              num_barbers=num_barbers,
56          )
57          for in_state, out_state in state_pairs
58      ]
59      transition_rates = np.reshape(
60          flat_transition_rates, (capacity + 1, capacity + 1)
61      )
62      np.fill_diagonal(
63          transition_rates, -transition_rates.sum(axis=1)
64      )
65
66      return transition_rates
```

Using this the matrix Q for the default system can be obtained:

Python input

```python
67  Q = get_transition_rate_matrix()
68  print(Q)
```

which gives:

```
Python output
69  [[-10  10   0   0   0   0   0]
70   [  4 -14  10   0   0   0   0]
71   [  0   8 -18  10   0   0   0]
72   [  0   0   8 -18  10   0   0]
73   [  0   0   0   8 -18  10   0]
74   [  0   0   0   0   8 -18  10]
75   [  0   0   0   0   0   8  -8]]
```

Here, the matrix exponential will be used as discussed above, using the SciPy [80] library. To see what would happen after 0.5 time units:

```
Python input
76  import scipy.linalg
77
78  print(scipy.linalg.expm(Q * 0.5).round(5))
```

which gives:

```
Python output
79  [[0.10492 0.21254 0.20377 0.17142 0.13021 0.09564 0.0815 ]
80   [0.08501 0.18292 0.18666 0.1708  0.14377 0.1189  0.11194]
81   [0.06521 0.14933 0.16338 0.16478 0.15633 0.14751 0.15346]
82   [0.04388 0.10931 0.13183 0.15181 0.16777 0.18398 0.21142]
83   [0.02667 0.07361 0.10005 0.13422 0.17393 0.2189  0.27262]
84   [0.01567 0.0487  0.07552 0.11775 0.17512 0.24484 0.32239]
85   [0.01068 0.03668 0.06286 0.10824 0.17448 0.25791 0.34914]]
```

To see what would happen after 500 time units:

```
Python input
86  print(scipy.linalg.expm(Q * 500).round(5))
```

which gives:

```
Python output
87  [[0.03431 0.08577 0.10722 0.13402 0.16752 0.2094  0.26176]
88   [0.03431 0.08577 0.10722 0.13402 0.16752 0.2094  0.26176]
89   [0.03431 0.08577 0.10722 0.13402 0.16752 0.2094  0.26176]
90   [0.03431 0.08577 0.10722 0.13402 0.16752 0.2094  0.26176]
91   [0.03431 0.08577 0.10722 0.13402 0.16752 0.2094  0.26176]
92   [0.03431 0.08577 0.10722 0.13402 0.16752 0.2094  0.26176]
93   [0.03431 0.08577 0.10722 0.13402 0.16752 0.2094  0.26176]]
```

After 500 time units, the probability of ending up in each state (column) is the same regardless of the state the system began in (row).

The analysis can in fact be stopped here however the choice of 500 time units was arbitrary and might not be the correct amount for all possible scenarios, as such the underlying equation 2.4 can be solved directly in order to get a solution where equilibrium is guaranteed.

The underlying linear system will be solved using a numerically efficient algorithm called least squares optimisation (available from the Numpy library):

```
Python input
94   def get_steady_state_vector(Q):
95       """Return the steady state vector of any given continuous time
96       transition rate matrix.
97
98       Args:
99           Q: a transition rate matrix
100
101      Returns:
102          A vector
103      """
104      state_space_size, _ = Q.shape
105      A = np.vstack((Q.T, np.ones(state_space_size)))
106      b = np.append(np.zeros(state_space_size), 1)
107      x, _, _, _ = np.linalg.lstsq(A, b, rcond=None)
108      return x
```

The steady state vector for the default system is given by:

```
Python input
109  print(get_steady_state_vector(Q).round(5))
```

giving:

Python output

```
110   [0.03431 0.08577 0.10722 0.13402 0.16752 0.2094  0.26176]
```

This shows that the shop is expected to be empty approximately 3.4% of the time and full 26.2% of the time.

The final function written is one that uses all of the above to return the probability of the shop being full.

Python input

```
111   def get_probability_of_full_shop(waiting_room=4, num_barbers=2):
112       """Return the probability of the barber shop being full.
113
114       Args:
115           waiting_room: an integer (default: 4)
116           num_barbers: an integer (default: 2)
117
118       Returns:
119           A real.
120       """
121       Q = get_transition_rate_matrix(
122           waiting_room=waiting_room,
123           num_barbers=num_barbers,
124       )
125       pi = get_steady_state_vector(Q)
126       return pi[-1]
```

This can now confirm the previous calculated probability of the shop being full:

Python input

```
127   print(round(get_probability_of_full_shop(), 6))
```

---

which gives:

Python output

```
128   0.261756
```

Now that the models have been defined, they will be used to compare the 2 possible scenarios.

Having 2 extra space in the waiting room corresponds to:

```
Python input
129  print(round(get_probability_of_full_shop(waiting_room=6), 6))
```

which gives:

```
Python output
130  0.23557
```

This is a slight improvement however, increasing the number of barbers has a substantial effect:

```
Python input
131  print(round(get_probability_of_full_shop(num_barbers=3), 6))
```

```
Python output
132  0.078636
```

Therefore, it would be better to increase the number of barbers by 1 than to increase the waiting room capacity by 2.

2.4 SOLVING WITH R

The first step taken is to write a function to obtain the transition rates between 2 given states:

```
R input
133  #' Return the transition rate for 2 given states.
134  #'
135  #' @param in_state an integer
136  #' @param out_state an integer
137  #' @param waiting_room an integer (default: 4)
138  #' @param num_barbers an integer  (default: 2)
139  #'
140  #' @return A real
141  get_transition_rate <- function(in_state,
142                                  out_state,
143                                  waiting_room = 4,
144                                  num_barbers = 2) {
```

```
145    arrival_rate <- 10
146    service_rate <- 4
147    capacity <- waiting_room + num_barbers
148    delta <- out_state - in_state
149
150    if (delta == 1) {
151      return(arrival_rate)
152    }
153    if (delta == -1) {
154      return(min(in_state, num_barbers) * service_rate)
155    }
156    return(0)
157  }
```

This actual function will not be used but instead a vectorized version[2] of this makes calculations more efficient:

R input

```
158  vectorized_get_transition_rate <- Vectorize(
159    get_transition_rate,
160    vectorize.args = c("in_state", "out_state")
161  )
```

This function can now take a vector of inputs for the `in_state` and `out_state` variables which will allow us to simplify the following code that creates the matrices:

R input

```
162  #' Return the transition rate matrix Q
163  #'
164  #' @param waiting_room an integer (default: 4)
165  #' @param num_barbers an integer (default: 2)
166  #'
167  #' @return A matrix
168  get_transition_rate_matrix <- function(waiting_room = 4,
169                                         num_barbers = 2){
170    max_state <- waiting_room + num_barbers
171
```

---

[2]A vectorized calculation refers to the manner in which an instruction is given to a computer. When vectorized: a single instruction with multiple data are given at the same time which corresponds to "Single instruction, multiple data" (SIMD) as defined in Flynn's taxonomy [19]. This is a type of parallelisation that can be done at the central processing unit level of the computer.

```
172   Q <- outer(
173       0:max_state,
174       0:max_state,
175       vectorized_get_transition_rate,
176       waiting_room = waiting_room,
177       num_barbers = num_barbers
178   )
179   row_sums <- rowSums(Q)
180   diag(Q) <- -row_sums
181   Q
182 }
```

Using this the matrix Q for the default system can be used:

R input

```
183   Q <- get_transition_rate_matrix()
184   print(Q)
```

which gives:

R output

```
185        [,1] [,2] [,3] [,4] [,5] [,6] [,7]
186   [1,]  -10   10    0    0    0    0    0
187   [2,]    4  -14   10    0    0    0    0
188   [3,]    0    8  -18   10    0    0    0
189   [4,]    0    0    8  -18   10    0    0
190   [5,]    0    0    0    8  -18   10    0
191   [6,]    0    0    0    0    8  -18   10
192   [7,]    0    0    0    0    0    8   -8
```

One immediate thing that can be done with this matrix is to take the matrix exponential discussed above. To do this, an R library called expm [25] will be used. To be able to make use of the nice %>% "pipe" operator the magrittr [2] library will be loaded. Now to see what would happen after 0.5 time units:

R input

```
193  library(expm)
194  library(magrittr)
195
196  print( (Q * 0.5) %>% expm() %>% round(5))
```

---

which gives:

R output

```
197          [,1]    [,2]    [,3]    [,4]    [,5]    [,6]    [,7]
198  [1,] 0.10492 0.21254 0.20377 0.17142 0.13021 0.09564 0.08150
199  [2,] 0.08501 0.18292 0.18666 0.17080 0.14377 0.11890 0.11194
200  [3,] 0.06521 0.14933 0.16338 0.16478 0.15633 0.14751 0.15346
201  [4,] 0.04388 0.10931 0.13183 0.15181 0.16777 0.18398 0.21142
202  [5,] 0.02667 0.07361 0.10005 0.13422 0.17393 0.21890 0.27262
203  [6,] 0.01567 0.04870 0.07552 0.11775 0.17512 0.24484 0.32239
204  [7,] 0.01068 0.03668 0.06286 0.10824 0.17448 0.25791 0.34914
```

After 500 time units:

R input

```
205  print( (Q * 500) %>% expm() %>% round(5))
```

---

which gives:

R output

```
206          [,1]    [,2]    [,3]    [,4]    [,5]   [,6]    [,7]
207  [1,] 0.03431 0.08577 0.10722 0.13402 0.16752 0.2094 0.26176
208  [2,] 0.03431 0.08577 0.10722 0.13402 0.16752 0.2094 0.26176
209  [3,] 0.03431 0.08577 0.10722 0.13402 0.16752 0.2094 0.26176
210  [4,] 0.03431 0.08577 0.10722 0.13402 0.16752 0.2094 0.26176
211  [5,] 0.03431 0.08577 0.10722 0.13402 0.16752 0.2094 0.26176
212  [6,] 0.03431 0.08577 0.10722 0.13402 0.16752 0.2094 0.26176
213  [7,] 0.03431 0.08577 0.10722 0.13402 0.16752 0.2094 0.26176
```

After 500 time units, the probability of ending up in each state (columns) is the same regardless of the state the system began in (row).

The analysis can in fact be stopped here however the choice of 500 time units was arbitrary and might not be the correct amount for all possible scenarios, as such the underlying equation 2.4 needs to be solved directly.

To be able to do this, the versatile pracma [3] package will be used which includes a number of numerical analysis functions for efficient computations.

```
R input
214  library(pracma)
215
216  #' Return the steady state vector of any given continuous time
217  #' transition rate matrix
218  #'
219  #' @param Q a transition rate matrix
220  #'
221  #' @return A vector
222  get_steady_state_vector <- function(Q){
223    state_space_size <- dim(Q)[1]
224    A <- rbind(t(Q), 1)
225    b <- c(integer(state_space_size), 1)
226    mldivide(A, b)
227  }
```

This is making use of pracma's `mldivide` function which chooses the best numerical algorithm to find the solution to a given matrix equation $Ax = b$.

The steady state vector for the default system is now given by:

```
R input
228  print(get_steady_state_vector(Q))
```

giving:

```
R output
229              [,1]
230  [1,] 0.03430888
231  [2,] 0.08577220
232  [3,] 0.10721525
233  [4,] 0.13401906
234  [5,] 0.16752383
235  [6,] 0.20940479
236  [7,] 0.26175598
```

The shop is expected to be empty approximately 3.4% of the time and full 26.2% of the time.

The final piece of this puzzle is to create a single function that uses all of the above to return the probability of the shop being full.

```
R input

237    #' Return the probability of the barber shop being full
238    #'
239    #' @param waiting_room (default: 4)
240    #' @param num_barbers (default: 2)
241    #'
242    #' @return A real
243    get_probability_of_full_shop <- function(waiting_room = 4,
244                                             num_barbers = 2){
245        arrival_rate <- 10
246        service_rate <- 4
247        pi <- get_transition_rate_matrix(
248          waiting_room = waiting_room,
249          num_barbers = num_barbers
250          ) %>%
251          get_steady_state_vector()
252
253        capacity <- waiting_room + num_barbers
254        pi[capacity + 1]
255    }
```

This confirms the previous calculated probability of the shop being full:

```
R input

256    print(get_probability_of_full_shop())
```

which gives:

```
R output

257    [1] 0.261756
```

Now that the models have been defined, they will be used to compare the 2 possible scenarios.

Adding 2 extra spaces in the waiting rooms corresponds to:

```
R input

258    print(get_probability_of_full_shop(waiting_room = 6))
```

which decreases the probability of a full shop to:

```
R output
259   [1] 0.2355699
```

but adding another barber and chair:

```
R input
260   print(get_probability_of_full_shop(num_barbers = 3))
```

gives:

```
R output
261   [1] 0.0786359
```

Therefore, it would be better to increase the number of barbers by 1 than to increase the waiting room capacity by 2.

2.5 WIDER CONTEXT

The overview of Markov chains given here has mainly concentrated on calculation of steady state probabilities. There are in fact many more theoretic and applied aspects of Markov chain models. Some examples of this include the calculation of sojourn times which is how long a system spends in a given state as well as considering models with absorption: where the system arrives at a state that it no longer leaves. For a good overview of these the following textbook is recommended: [67].

In [65, 70], Markov chains are used to model board games. In [70], a model of the battles that take place on a Risk board is used to understand the probabilities of invasion of territories based on troop numbers. This is done using an absorbing Markov chain. In [65], a standard model is used to identify the properties that are most likely to be landed on in Monopoly. This is done through calculation of steady state probabilities. These are both examples of discrete time Markov chains.

A common application of Markov chains is in queueing systems and specifically queueing systems applied to healthcare. In [26], a model of a neurological rehabilitation unit is built and used to help better staff the unit. This is accomplished using the steady state probabilities and calculating various performance measures. This is an application of a continuous time Markov chain.

An extension of Markov chains are Markov decision processes. This is a particular mathematical model that identifies the optimal decision made within a Markov chain. Instead of building multiple Markov models for different decisions, in Markov decision processes decisions can be made at each state of the underlying chain. A policy can be identified giving the optimal decision at each state. In [82], a literature review is given showing a wide ranging application of these decision processes from agriculture to motor insurance claims as well as sports.

CHAPTER 3

Discrete Event Simulation

C OMPLEX situations further compounded by randomness appear throughout daily lives. Examples include data flowing through a computer network, patients being treated at emergency services, and daily commutes to work. Mathematics can be used to understand these complex situations so as to make predictions, which in turn can be used to make improvements. One tool for this is to let a computer create a dynamic virtual representation of the scenario in question, a particular approach we are going to cover here is called discrete event simulation.

3.1 PROBLEM

A bicycle repair shop would like to reconfigure in order to guarantee that all bicycles processed take a maximum of 30 minutes. Their current set-up is as follows:

- Bicycles arrive randomly at the shop at a rate of 15 per hour;

- they wait in line to be seen at an inspection counter, staffed by one member of staff who can inspect one bicycle at a time. On average an inspection takes around 3 minutes;

- around 20% of bicycles do not need repair after inspection, and they are then ready for collection;

- around 80% of bicycles go on to be repaired after inspection. These then wait in line outside the repair workshop, which is staffed by two members of staff who can each repair one bicycle at a time. On average a repair takes around 6 minutes;

- after repair the bicycles are ready for collection.

A diagram of the system is shown in Figure 3.1.

An assumption of infinite capacity at the bicycle repair shop for waiting bicycles is made. The shop will hire an extra member of staff in order to meet their target of a maximum time in the system of 30 minutes. They would like to know if they should work on the inspection counter or in the repair workshop.

DOI: 10.1201/9780429328534-3

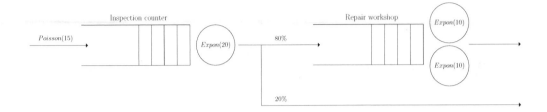

Figure 3.1 Diagrammatic representation of the bicycle repair shop as a queuing system.

3.2 THEORY

A number of aspects of the bicycle shop above are probabilistic. For example, the times that bicycles arrive at the shop, the duration of the inspection and repairs, and whether the bicycle would need to go on to be repaired or not. When a number of these probabilistic events are linked together such as the bicycle shop, a method to model this situation is discrete event simulation (DES).

Consider one probabilistic event, rolling a six-sided die where each side is equally likely to land. Therefore the probability of rolling a 1 is $\frac{1}{6}$, the probability of rolling a 2 is $\frac{1}{6}$, and so on. This means that if the die is rolled a large number of times, $\frac{1}{6}$ of those rolls would be expected to be a 1.

Consider a random process in which the actual values of the probability of events occurring are not known. Consider rolling a weighted die, in this case a die in which the probability of obtaining one number is much greater than the others. How can probability of obtaining a 1 on this die be estimated?

Rolling the weighted die once does not give much information. However due to a theorem called the law of large numbers [67], this die can be rolled a number of times and find the proportion of those rolls which gave a 1. The more times we roll the die, the closer this proportion approaches the actual value of the probability of obtaining a 1.

For a complex system such as the bicycle shop, the goal is to estimate the proportion of bicycles that take longer than 30 minutes to be processed. As it is a complex system it is difficult to obtain an exact value. So, like the weighted die, the system will be observed a number of times and the overall proportions of bicycles spending longer than 30 minutes in the shop will converge to the exact value. Unlike rolling a weighted die, it is costly to observe this shop over a number of days with identical conditions. In this case, it is costly in terms of time, as the repair shop already exists. However some scenarios, for example the scenario where the repair shop hires an additional member of staff, do not yet exist, so observing this would be costly in terms of money also. It is possible to build a virtual representation of this complex system on a computer, and observe a virtual day of work much more quickly and with much less cost, similar to a video game.

In order to do this, the computer needs to be able to generate random outcomes of each of the smaller events that make up the large complex system. Generating random events corresponds to sampling random numbers.

Computers are deterministic; therefore true randomness is in itself a challenging mathematical problem. They can however generate pseudorandom numbers: sequences of numbers that look like random numbers, but are entirely determined from the previous numbers in the sequence[1]. Most programming languages have methods of doing this.

In order to simulate an event the law of large numbers can be used. Let $X \sim U(0, 1)$, a uniformly pseudorandom variable between 0 and 1. Let D be the outcome of a roll of an unbiased die. Then D can be defined as:

$$D = \begin{cases} 1 & \text{if } 0 \leq X < \frac{1}{6} \\ 2 & \text{if } \frac{1}{6} \leq X < \frac{2}{6} \\ 3 & \text{if } \frac{2}{6} \leq X < \frac{3}{6} \\ 4 & \text{if } \frac{3}{6} \leq X < \frac{4}{6} \\ 5 & \text{if } \frac{4}{6} \leq X < \frac{5}{6} \\ 6 & \text{if } \frac{5}{6} \leq X < 1 \end{cases} \tag{3.1}$$

The bicycle repair shop is a system of interactions of random events. This can be thought of as many interactions of random variables, each generated using pseudorandom numbers.

In this case, the fundamental random events that need to be generated are:

- the time each bicycle arrives to the repair shop,

- the time each bicycle spends at the inspection counter,

- whether each bicycle needs to go on to the repair workshop,

- the time those bicycles spend being repaired.

As the simulation progresses these events will be generated, and will interact together as described in Section 3.1. The proportion of customers spending longer than 30 minutes in the shop can then be counted. This proportion itself is a random variable, and like the weighted die, running this simulation once does not give much information. The simulation can be run many times to give an average proportion.

The process outlined above is a particular implementation of Monte Carlo simulation called discrete event simulation, which is a generic term for generating pseudorandom numbers and observes the emergent interactions. In practice there are two main approaches to simulating complex probabilistic systems such as this one: *event scheduling* and *process-based* simulation. It so happens that the implementations in Python and R shown here use each of these approaches respectively.

---

[1]An early discussion of pseudorandom numbers is [81]. A number of different pseudorandom number generators exist, at the time of writing the state of the art is the Mersenne Twister described in [39].

3.2.1 Event Scheduling Approach

When using the event scheduling approach, the "virtual representation" of the system is the collection of facilities that the bicycles use, shown in Figure 3.1. Then the entities (the bicycles) interact with these facilities. It is these facilities that determine how the entities behave.

In a simulation that uses an event scheduling approach, a key concept is that when events occur, this causes further events to occur in the future, either immediately or after a delay. In the bicycle shop, examples of such events include a bicycle joining a queue, a bicycle beginning service and a bicycle finishing service. At each event the event list is updated, and the clock then jumps forward to the next event in this updated list.

3.2.2 Process-Based Simulation

When using process-based simulation, the "virtual representation" of the system is the sequence of actions that each entity (the bicycles) must take, and these sequences of actions might contain delays as a number of entities seize and release a finite amount of resources. It is the sequence of these actions that determine how the entities behave.

For the bicycle repair shop, an example of one possible sequence of actions would be:

arrive → seize inspection counter → delay → release inspection counter → seize repair shop → delay → release repair shop → leave

The scheduled delays in this sequence of events correspond to the time spent being inspected and the time spent being repaired. Waiting in line for service at these facilities is not included in the sequence of events; that is implicit by the "seize" and "release" actions, as an entity will wait for a free resource before seizing one. Therefore in process-based simulations, in addition to defining a sequence of events, resource types and their numbers also need to be defined.

3.3 SOLVING WITH PYTHON

In this book, the Ciw [52, 72] library will be used in order to conduct discrete event simulation in Python. Ciw uses the event scheduling approach, which means the system's facilities are defined, and customers then interact with them.

In this case, there are two facilities to define: the inspection desk and the repair workshop. For each of these the following need to be defined:

- the distribution of times between consecutive bicycles arriving,

- the distribution of times the bicycles spend in service,

- the number of servers available,

- the probability of routing to each of the other facilities after service.

In this case, the time between consecutive arrivals is assumed to follow an exponential distribution, as is the service time. These are common assumptions for this sort of queueing system [67].

In Ciw, these are defined as part of a Network object, created using the `ciw.create_network` function. The function below creates a Network object that defines the system for a given set of parameters bicycle repair shop:

```
Python input

262  import ciw
263
264
265  def build_network_object(
266      num_inspectors=1,
267      num_repairers=2,
268  ):
269      """Returns a Network object that defines the repair shop.
270
271      Args:
272          num_inspectors: a positive integer (default: 1)
273          num_repairers: a positive integer (default: 2)
274
275      Returns:
276          a Ciw network object
277      """
278      arrival_rate = 15
279      inspection_rate = 20
280      repair_rate = 10
281      prob_need_repair = 0.8
282      N = ciw.create_network(
283          arrival_distributions=[
284              ciw.dists.Exponential(arrival_rate),
285              ciw.dists.NoArrivals(),
286          ],
287          service_distributions=[
288              ciw.dists.Exponential(inspection_rate),
289              ciw.dists.Exponential(repair_rate),
290          ],
291          number_of_servers=[num_inspectors, num_repairers],
292          routing=[[0.0, prob_need_repair], [0.0, 0.0]],
293      )
294      return N
```

A Network object is used by Ciw to access system parameters. For example, one piece of information it holds is the number of nodes of the system:

```
Python input

295  N = build_network_object()
296  print(N.number_of_nodes)
```

which gives:

```
Python output

297  2
```

Now that the system is defined a Simulation object can be created. Once this is built the simulation can be run, that is observe it for one virtual day. The following function does this:

```
Python input

298  def run_simulation(network, seed=0):
299      """Builds a simulation object and runs it for 8 time units.
300
301      Args:
302          network: a Ciw network object
303          seed: a float (default: 0)
304
305      Returns:
306          a Ciw simulation object after a run of the simulation
307      """
308      max_time = 8
309      ciw.seed(seed)
310      Q = ciw.Simulation(network)
311      Q.simulate_until_max_time(max_time)
312      return Q
```

Notice here that a random seed is set. This is because there is randomness in running the simulation, setting a seed ensures reproducible results[2]. Notice also that the simulation always begins with an empty system, so the first bicycle to arrive will never wait for service. Depending on the situation this may be an unwanted feature, though not in this case as it is reasonable to assume that the bicycle shop will begin the day with no customers.

---

[2]Pseudorandom number generators give a sequence of numbers that obey a series of properties. A seed is necessary to obtain a starting point for a given sequence. This has the benefit of ensuring that given sequences can be reproduced.

To count the number of bicycles that have finished service, and to count the number of those whose entire journey through the system lasted longer than 0.5 hours the Pandas [40, 71] library will be used:

```
Python input

313  import pandas as pd
314
315
316  def get_proportion(Q):
317      """Returns the proportion of bicycles spending over a given
318      limit at the repair shop.
319
320      Args:
321          Q: a Ciw simulation object after a run of the
322              simulation
323
324      Returns:
325          a real
326      """
327      limit = 0.5
328      inds = Q.nodes[-1].all_individuals
329      recs = pd.DataFrame(
330          dr for ind in inds for dr in ind.data_records
331      )
332      recs["total_time"] = recs["exit_date"] - recs["arrival_date"]
333      total_times = recs.groupby("id_number")["total_time"].sum()
334      return (total_times > limit).mean()
```

Altogether these functions can define the system, run one day of the system, and then find the proportion of bicycles spending over half an hour in the shop:

```
Python input

335  N = build_network_object()
336  Q = run_simulation(N)
337  p = get_proportion(Q)
338  print(round(p, 6))
```

This gives:

```
Python output

339  0.261261
```

meaning 26.13% of all bicycles spent longer than half an hour at the repair shop.

However this particular day may have contained a number of extreme events. For a more accurate proportion this experiment should be repeated a number of times, and an average proportion taken. The following function returns an average proportion:

Python input

```python
340  def get_average_proportion(num_inspectors=1, num_repairers=2):
341      """Returns the average proportion of bicycles spending over a
342      given limit at the repair shop.
343
344      Args:
345          num_inspectors: a positive integer (default: 1)
346          num_repairers: a positive integer (default: 2)
347
348      Returns:
349          a real
350      """
351      num_trials = 100
352      N = build_network_object(
353          num_inspectors=num_inspectors,
354          num_repairers=num_repairers,
355      )
356      proportions = []
357      for trial in range(num_trials):
358          Q = run_simulation(N, seed=trial)
359          proportion = get_proportion(Q=Q)
360          proportions.append(proportion)
361      return sum(proportions) / num_trials
```

This can be used to find the average proportion over 100 trials for the current system of one inspector and two repair people:

Python input

```python
362  p = get_average_proportion(num_inspectors=1, num_repairers=2)
363  print(round(p, 6))
```

which gives:

Python output

```
364  0.159354
```

that is, on average 15.94% of bicycles will spend longer than 30 minutes at the repair shop.

Now consider the two possible future scenarios: hiring an extra member of staff to serve at the inspection desk, or hiring an extra member of staff at the repair workshop. Which scenario yields a smaller proportion of bicycles spending over 30 minutes at the shop? First look at the situation where the additional member of staff works at the inspection desk is considered:

Python input

```
365  p = get_average_proportion(num_inspectors=2, num_repairers=2)
366  print(round(p, 6))
```

which gives:

Python output

```
367  0.038477
```

that is 3.85% of bicycles.

Now look at the situation where the additional member of staff works at the repair workshop:

Python input

```
368  p = get_average_proportion(num_inspectors=1, num_repairers=3)
369  print(round(p, 6))
```

which gives:

Python output

```
370  0.103591
```

that is 10.36% of bicycles.

Therefore an additional member of staff at the inspection desk would be more beneficial than an additional member of staff at the repair workshop.

3.4   SOLVING WITH R

In this book the Simmer [76] package will be used in order to conduct discrete event simulation in R. Simmer uses the process-based approach, which means that each bicycle's sequence of actions must be defined, and then generate a number of bicycles with these sequences.

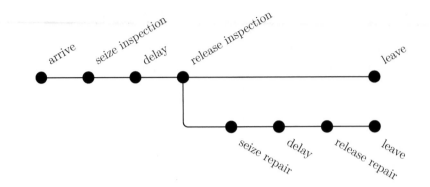

Figure 3.2 Diagrammatic representation of the forked trajectories a bicycle can take.

In Simmer these sequences of actions are made up of trajectories. The diagram in Figure 3.2 shows the branched trajectories than a bicycle would take at the repair shop:

The function below defines a simmer object that describes these trajectories:

```
R input

371  library(simmer)
372
373  #' Returns a simmer trajectory object outlining the bicycles
374  #' path through the repair shop
375  #'
376  #' @return A simmer trajectory object
377  define_bicycle_trajectories <- function() {
378    inspection_rate <- 20
379    repair_rate <- 10
380    prob_need_repair <- 0.8
381    bicycle <-
382      trajectory("Inspection") %>%
383      seize("Inspector") %>%
384      timeout(function() {
385        rexp(1, inspection_rate)
386      }) %>%
387      release("Inspector") %>%
388      branch(
389        function() (runif(1) < prob_need_repair),
390        continue = c(F),
391        trajectory("Repair") %>%
392          seize("Repairer") %>%
393          timeout(function() {
```

```
394          rexp(1, repair_rate)
395        }) %>%
396        release("Repairer"),
397      trajectory("Out")
398    )
399    return(bicycle)
400  }
```

These trajectories are not useful alone as the resources used are yet to be defined, or a way to generate bicycles with these trajectories. This is done in the function below, which begins by defining a **repair_shop** with one resource labelled "Inspector", and two resources labelled "Repairer". Once this is built the simulation can be run, that is observe it for one virtual day. The following function does all this:

R input

```
401  #' Runs one trial of the simulation.
402  #'
403  #' @param bicycle a simmer trajectory object
404  #' @param num_inspectors positive integer (default: 1)
405  #' @param num_repairers positive integer (default: 2)
406  #' @param seed a float (default: 0)
407  #'
408  #' @return A simmer simulation object after one run of
409  #'         the simulation
410  run_simulation <- function(bicycle,
411                             num_inspectors = 1,
412                             num_repairers = 2,
413                             seed = 0) {
414    arrival_rate <- 15
415    max_time <- 8
416    repair_shop <-
417      simmer("Repair Shop") %>%
418      add_resource("Inspector", num_inspectors) %>%
419      add_resource("Repairer", num_repairers) %>%
420      add_generator(
421        "Bicycle", bicycle, function() {
422          rexp(1, arrival_rate)
423        }
424      )
425    set.seed(seed)
426    repair_shop %>% run(until = max_time)
```

```
427      return(repair_shop)
428  }
```

Notice here a random seed is set. This is because there are elements of randomness when running the simulation, setting a seed ensures reproducible results[3]. Notice also that the simulation always begins with an empty system, so the first bicycle to arrive will never wait for service. Depending on the situation this may be an unwanted feature, though not in this case as it is reasonable to assume that the bicycle shop will begin the day with no customers.

To count the number of bicycles that have finished service, and to count the number of those whose entire journey through the system lasted longer than 0.5 hours, Simmer's `get_mon_arrivals()` function gives a data frame that can be manipulated:

R input

```
429  #' Returns the proportion of bicycles spending over 30
430  #' minutes in the repair shop
431  #'
432  #' @param repair_shop a simmer simulation object
433  #'
434  #' @return a float between 0 and 1
435  get_proportion <- function(repair_shop) {
436      limit <- 0.5
437      recs <- repair_shop %>% get_mon_arrivals()
438      total_times <- recs$end_time - recs$start_time
439      return(mean(total_times > limit))
440  }
```

Altogether these functions can define the system, run one day of the system and then find the proportion of bicycles spending over half an hour in the shop:

R input

```
441  bicycle <- define_bicycle_trajectories()
442  repair_shop <- run_simulation(bicycle = bicycle)
443  print(get_proportion(repair_shop = repair_shop))
```

This piece of code gives

---

[3]Pseudorandom number generators give a sequence of numbers that obey a series of properties. A seed is necessary to obtain a starting point for a given sequence. This has the benefit of ensuring that given sequences can be reproduced.

```
 R output

444  [1] 0.1343284
```

meaning 13.43% of all bicycles spent longer than half an hour at the repair shop.
However, this particular day may have contained a number of extreme events.
For a more accurate proportion this experiment should be repeated a number of
times, and an average proportion taken. In order to do so, the following is a function
that performs the above experiment over a number of trials, then finds an average
proportion:

```
 R input

445  #' Returns the average proportion of bicycles spending over
446  #' a given limit at the repair shop.
447  #'
448  #' @param num_inspectors positive integer (default: 1)
449  #' @param num_repairers positive integer (default: 2)
450
451  #' @return a float between 0 and 1
452  get_average_proportion <- function(num_inspectors = 1,
453                                     num_repairers = 2) {
454    num_trials <- 100
455    bicycle <- define_bicycle_trajectories()
456    proportions <- c()
457    for (trial in 1:num_trials) {
458      repair_shop <- run_simulation(
459        bicycle = bicycle,
460        num_inspectors = num_inspectors,
461        num_repairers = num_repairers,
462        seed = trial
463      )
464      proportion <- get_proportion(
465        repair_shop = repair_shop
466      )
467      proportions[trial] <- proportion
468    }
469    return(mean(proportions))
470  }
```

This can be used to find the average proportion over 100 trials:

```
R input
471  print(
472    get_average_proportion(
473      num_inspectors = 1,
474      num_repairers = 2)
475  )
```

which gives:

```
R output
476  [1] 0.1635779
```

that is, on average 16.36% of bicycles will spend longer than 30 minutes at the repair shop.

Now consider the two possible future scenarios: hiring an extra member of staff to serve at the inspection desk, or hiring an extra member of staff at the repair workshop. Which scenario yields a smaller proportion of bicycles spending over 30 minutes at the shop? First consider the situation where the additional member of staff works at the inspection desk:

```
R input
477  print(
478    get_average_proportion(
479      num_inspectors = 2,
480      num_repairers = 2
481    )
482  )
```

which gives:

```
R output
483  [1] 0.04221602
```

that is 4.22% of bicycles.

Now look at the situation where the additional member of staff works at the repair workshop:

```
R input
484  print(
485    get_average_proportion(
486      num_inspectors = 1,
487      num_repairers = 3
488    )
489  )
```

which gives:

```
R output
490  [1] 0.1224761
```

that is 12.25% of bicycles.

Therefore an additional member of staff at the inspection desk would be more beneficial than an additional member of staff at the repair workshop.

3.5 WIDER CONTEXT

The concepts shown in this chapter cover some theoretical aspects of discrete event simulation. There are a number of further topics that can be vital to creating valid models of real-life scenarios. These include time-dependent rates and rostering servers. An overview of the theory of discrete event simulation is given in [56].

One particular use of discrete event simulation is as part of a wider optimisation exercise. For example, a systematic search for an optimal service configuration can use a discrete event simulation model as a replacement for a mathematical objective function. Another approach is to integrate an optimisation procedure[4] within a discrete event simulation model so as to iteratively simulate optimal configurations. This is done in [51] to be able to bring together strategic configuration and overall flow in the blood supply chain. A general review and taxonomy of different uses of discrete event simulation with optimisation techniques is given in [18].

One domain where simulation is often used is in modelling healthcare systems. A general overview is given in [6] where uses include resource utilisation, human behaviour and workforce management.

In order to be able to fully capture all the relevant details of the system to be modelled, an extension of discrete event simulation is to combine the methodology with systems dynamics (Chapter 5) in order to model continuous aspects of the system and/or agent-based modelling (Chapter 7) in order to observe emergent or learned behaviours. This is known as hybrid simulation, and an overview is given in [7].

---

[4]For more information on optimisation see Chapters 8 and 9.

III

Dynamical Systems

CHAPTER 4

Differential Equations

S YSTEMS often change in a way that depends on their current state. For example, the speed at which a cup of coffee cools down depends on its current temperature. These types of systems are called dynamical systems and are modelled mathematically using differential equations. This chapter will consider a direct solution approach using symbolic mathematics.

4.1 PROBLEM

Consider the following situation: the entire population of a small rural town has caught a cold. All 100 individuals will recover at an average rate of 2 per day. The town leadership have noticed that being ill costs approximately £10 per day, this is due to general lack of productivity, poorer mood and other intangible aspects. They need to decide whether or not to order cold medicine, which would **double** the recovery rate. The cost of the cold medicine is a one off cost of £5 per person.

4.2 THEORY

In the case of this town, the overall rate at which people get better is dependent on the number of people who are ill. This can be represented mathematically using a differential equation, which is a way of relating the rate of change of a system to the state of the system itself.

In general the objects of interest are the variable x over time t, and the rate at which x changes with t, its derivative $\frac{dx}{dt}$. The differential equation describing this will be of the form:

$$\frac{dx}{dt} = f(x) \tag{4.1}$$

for some function f. In this case, the number of infected individuals will be denoted as I, which will implicitly mean that I is a function of time: $I = I(t)$, and the rate at which individuals recover will be denoted by α, then the differential equation that describes the above situation is:

$$\frac{dI}{dt} = -\alpha I \tag{4.2}$$

DOI: 10.1201/9780429328534-4

Finding a solution to this differential equation means finding an expression for I that when differentiated gives $-\alpha I$.

In this particular case, one such function is:

$$I(t) = e^{-\alpha t} \qquad (4.3)$$

This is a solution because: $\frac{dI}{dt} = -\alpha e^{-\alpha y} = -\alpha I$.

However here $I(0) = 1$, whereas for this problem we know that at time $t = 0$ there are 100 infected individuals. In general there are many such functions that can satisfy a differential equation, known as a family of solutions. To know which particular solution is relevant to the situation, some sort of initial condition is required. Here this would be:

$$I(t) = 100e^{-\alpha t} \qquad (4.4)$$

To evaluate the cost the sum of the values of that function over time is needed. Integration gives exactly this, so the cost would be:

$$K \int_0^\infty I(t)dt \qquad (4.5)$$

where K is the cost per person per unit time. Therefore the overall cost would be the cost of being unproductive, plus the cost of the medicine, M, and is given by:

$$K \int_0^\infty I(t)dt + MI(0) \qquad (4.6)$$

In the upcoming sections code will be used to confirm to carry out the above efficiently so as to answer the original question.

4.3 SOLVING WITH PYTHON

The first step is to define the symbolic variables that will be used. The Python library SymPy [42] is used.

```
Python input

491   import sympy as sym
492
493   t = sym.Symbol("t")
494   alpha = sym.Symbol("alpha")
495   I_0 = sym.Symbol("I_0")
496   I = sym.Function("I")
```

Now write a function to obtain the differential equation.

```
     Python input
497  def get_equation(alpha=alpha):
498      """Return the differential equation.
499
500      Args:
501          alpha: a float (default: symbolic alpha)
502
503      Returns:
504          A symbolic equation
505      """
506      return sym.Eq(sym.Derivative(I(t), t), -alpha * I(t))
```

This gives an equation that defines the population change over time:

```
     Python input
507  eq = get_equation()
508  print(eq)
```

which gives:

```
     Python output
509  Eq(Derivative(I(t), t), -alpha*I(t))
```

Note that if Jupyter [33] notebooks are used, then output will actually be a well rendered mathematical equation:

$$\frac{d}{dt}I(t) = -\alpha I(t)$$

A value of α can be passed if required:

```
     Python input
510  eq = get_equation(alpha=1)
511  print(eq)
```

```
     Python output
512  Eq(Derivative(I(t), t), -I(t))
```

Now a function will be written to obtain the solution to this differential equation with initial condition $I(0) = I_0$:

```
Python input
513  def get_solution(I_0=I_0, alpha=alpha):
514      """Return the solution to the differential equation.
515
516      Args:
517          I_0: a float (default: symbolic I_0)
518          alpha: a float (default: symbolic alpha)
519
520      Returns:
521          A symbolic equation
522      """
523      eq = get_equation(alpha=alpha)
524      return sym.dsolve(eq, I(t), ics={I(0): I_0})
```

This can verify the solution discussed previously:

```
Python input
525  sol = get_solution()
526  print(sol)
```

which gives:

```
Python output
527  Eq(I(t), I_0*exp(-alpha*t))
```

$$I(t) = I_0 e^{-\alpha t}$$

SymPy itself can be used to verify the result, by taking the derivative of the right-hand side of our solution.

```
Python input
528  print(sym.diff(sol.rhs, t) == -alpha * sol.rhs)
```

which gives:

Python output

```
529  True
```

All of the above have given the general solution in terms of $I(0) = I_0$ and α; however, the code is written in such a way as we can pass the actual parameters:

Python input

```
530  sol = get_solution(alpha=2, I_0=100)
531  print(sol)
```

which gives:

Python output

```
532  Eq(I(t), 100*exp(-2*t))
```

Now, to calculate the cost write a function to integrate the result:

Python input

```
533  def get_cost(
534      I_0=I_0,
535      alpha=alpha,
536      per_person_cost=10,
537      cure_cost=0,
538  ):
539      """Return the cost.
540
541      Args:
542          I_0: a float (default: symbolic I_0)
543          alpha: a float (default: symbolic alpha)
544          per_person_cost: a float (default: 10)
545          cure_cost: a float (default: 0)
546
547      Returns:
548          A symbolic expression
549      """
550      I_sol = get_solution(I_0=I_0, alpha=alpha)
551      area = sym.integrate(I_sol.rhs, (t, 0, sym.oo))
552      productivity_cost = area * per_person_cost
```

```
553        total_cost_of_cure = cure_cost * I_0
554        return productivity_cost + total_cost_of_cure
```

The cost without purchasing the cure is:

Python input

```
555  alpha = 2
556  cost_without_cure = get_cost(I_0=100, alpha=alpha)
557  print(cost_without_cure)
```

which gives:

Python output

```
558  500
```

The cost with cure can use the above with a modified α and a non-zero cost of the cure itself:

Python input

```
559  cost_with_cure = get_cost(I_0=100, alpha=2 * alpha, cure_cost=5)
560  print(cost_with_cure)
```

which gives:

Python output

```
561  750
```

So given the current parameters it is not worth purchasing the cure.

4.4 SOLVING WITH R

R has some capability for symbolic mathematics; however at the time of writing, the options available are somewhat limited and/or not reliable. As such, in R, the problem will be solved using a numerical integration approach. For an outline of the theory behind this approach see Chapter 5.

First a function to give the derivative for a given value of I is needed:

```
R input

562  #' Returns the numerical value of the derivative.
563  #'
564  #' @param t a set of time points
565  #' @param y a function
566  #' @param parameters the set of all parameters passed to y
567
568  #' @return a float
569  derivative <- function(t, y, parameters) {
570    with(
571      as.list(c(y, parameters)), {
572        dIdt <- -alpha * I
573        list(dIdt)
574      }
575    )
576  }
```

For example, to see the value of the derivative when $I = 0$:

```
R input

577  dv <- derivative(t = 0, y = c(I = 100), parameters = c(alpha = 2))
578  print(dv)
```

This gives:

```
R output

579  [[1]]
580  [1] -200
```

Now the deSolve [64] library will be used for solving differential equations numerically:

```
R input

581  library(deSolve)
582  #' Return the solution to the differential equation.
583  #'
584  #' @param times: a vector of time points
585  #' @param y_0: a float (default: 100)
586  #' @param alpha: a float (default: 2)
```

```
587
588    #' @return A vector of numerical values
589    get_solution <- function(times,
590                             y0 = c(I = 100),
591                             alpha = 2) {
592      params <- c(alpha = alpha)
593      ode(y = y0, times = times, func = derivative, parms = params)
594    }
```

This will return a sequence of time point and values of I at those time points. Using this we can compute the cost.

R input

```
595    #' Return the cost.
596    #'
597    #' @param I_0: a float (default: symbolic I_0)
598    #' @param alpha: a float (default: symbolic alpha)
599    #' @param per_person_cost: a float (default: 10)
600    #' @param cure_cost: a float (default: 0)
601    #' @param step_size: a float (default: 0.0001)
602    #' @param max_time: an integer (default: 10)
603
604    #' @return A numeric value
605    get_cost <- function(I_0 = 100,
606                         alpha = 2,
607                         per_person_cost = 10,
608                         cure_cost = 0,
609                         step_size = 0.0001,
610                         max_time = 10) {
611      times <- seq(0, max_time, by = step_size)
612      out <- get_solution(times, y0 = c(I = I_0), alpha = alpha)
613      number_of_observations <- length(out[, "I"])
614      time_intervals <- diff(out[, "time"])
615      area <- sum(time_intervals * out[-number_of_observations, "I"])
616      productivity_cost <- area * per_person_cost
617      total_cost_of_cure <- cure_cost * I_0
618      productivity_cost + total_cost_of_cure
619    }
```

The cost without purchasing the cure is:

```
R input

620  alpha <- 2
621  cost_without_cure <- get_cost(alpha = alpha)
622  print(round(cost_without_cure))
```

which gives:

```
R output

623  [1] 500
```

The cost with cure can use the above with a modified α and a non-zero cost of the cure itself:

```
R input

624  cost_with_cure <- get_cost(alpha = 2 * alpha, cure_cost = 5)
625  print(round(cost_with_cure))
```

which gives:

```
R output

626  [1] 750
```

So given the current parameters it is not worth purchasing the cure.

4.5   WIDER CONTEXT

There are a number of further areas related to the study of as well as the use of differential equations. Topics omitted here include the actual solution approaches which in this chapter are taken care of using open-source software. Chapters 9, 14 and 16 of [66] provide a good introduction to some of these concepts as well as a general discussion of the area of mathematics in which they sit: .

Differential equations have been applied in many settings. In [35] differential equations were used to model attrition in warfare, the insights from these differential equations are referred to as Lanchester's square law. This has been historically fitted to a number of battles with varying levels of success.

In [69] differential equations are used to build a generic model of regime change. A detailed analysis of the stability of the system is included. The model offers some explanation of why oppressive regimes can follow an overthrow of a similarly oppressive regime: the underlying mathematical system is a stable cycle from which it is difficult to escape.

[78] uses differential equations as a framework for modelling queueing networks. This is interesting in its inception as differential equations are models for continuous quantities, whereas queues are discrete-type events (see Chapters 2 and 3 for more on this). The advantages of using differential equations are mainly in the computational efficiency.

The model presented in this chapter is deterministic: there is a single solution that remains the same. This is not always a precise model of reality: often systems are stochastic so that the inputs are not constant parameters but follow some random distribution. This is where stochastic differential equations are applied, which is the subject of [58].

Systems Dynamics

I N many situations systems are dynamical, in that the state or population of a number of entities or classes change according to the current state or population of the system. For example population dynamics, chemical reactions, and macroeconomic systems. It is often useful to be able to predict how these systems will behave over time, though the rules that govern these changes may be complex, and are not necessarily solvable analytically. In these cases numerical methods may be needed; this is the focus of this chapter.

5.1 PROBLEM

Consider the following scenario, where a population of 3000 people are susceptible to infection by some disease. This population can be described by the following parameters:

- They have a birth rate b of 0.01 per day;

- they have a death rate d of 0.01 per day;

- for every infectious individual, the infection rate α is 0.3 per day;

- infectious people recover naturally and gain an immunity from the disease, at a recovery rate r of 0.02 per day;

- for each day an individual is infected, they must take medication which costs a public healthcare system £10 per day.

A vaccine is produced, that allows individuals to gain an immunity. This vaccine costs the public healthcare system a one-off cost of £220 per vaccine. The healthcare providers would like to know if achieving a vaccination rate v of 85% would be beneficial financially.

5.2 THEORY

The above scenario can be expressed using a compartmental model of disease, and can be represented in a stock and flow diagram as in Figure 5.1.

DOI: 10.1201/9780429328534-5

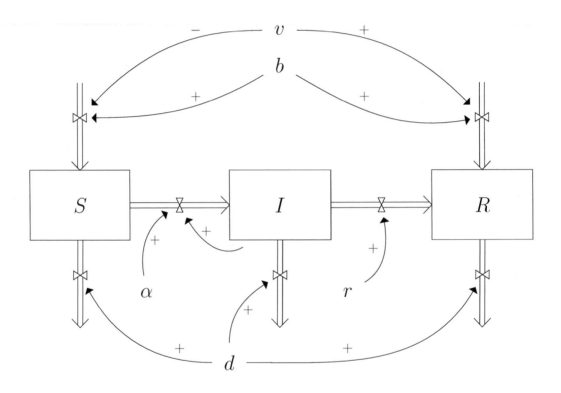

Figure 5.1 Diagrammatic representation of the epidemiology model.

The system has three quantities, or "stocks", of different types of individuals, those susceptible to disease (S), those infected with the disease (I) and those who have recovered from the disease and so have gained immunity (R)[1]. The levels on these stocks change according to the flows in, out, and between them, controlled by "taps". The amount of flow the taps let through are influenced in a multiplicative way (either negatively or positively), by other factors, such as external parameters (e.g. birth rate, infection rate) and the stock levels.

In this system the following taps exist, influenced by the following parameters:

- *external → S:* influenced positively by the birth rate, and negatively by the vaccine rate;

- *S → I:* influenced positively by the infection rate, and the number of infected individuals;

- *S → external:* influenced positively by the death rate;

- *I → R:* influenced positively by the recovery rate;

- *I → external:* influenced positively by the death rate;

- *R → external:* influenced positively by the birth rate and the vaccine rate;

- *external → R:* influenced positively by the death rate.

Mathematically the quantities or stocks are functions over time, and the changes in stock levels are written as their derivatives, for example the change in the number of susceptible individuals over time is denoted by $\frac{dS}{dt}$. This is equal to the sum of the taps in or out of that stock. Thus the system is described by the following system of differential equations:

$$\frac{dS}{dt} = -\frac{\alpha SI}{N} + (1-v)bN - dS \tag{5.1}$$

$$\frac{dI}{dt} = \frac{\alpha SI}{N} - (r+d)I \tag{5.2}$$

$$\frac{dR}{dt} = rI - dR + vbN \tag{5.3}$$

where $N = S + I + R$ is the total number of individuals in the system.

The behaviour of the quantities S, I and R under these rules can be quantified by solving this system of differential equations. This system contains some non-linear terms, implying that this may be difficult to solve analytically, so a numerical method instead will be used.

A number of potential numerical methods to do this exist. The solvers that will be used in Python and R choose the most appropriate for the problem at hand. In general methods for this kind of problems use the principle that the derivative denotes the

---

[1]This is often referred to as an SIR model.

rate of instantaneous change. Thus for a differential equation $\frac{dy}{dt} = f(t, y)$, consider the function y as a discrete sequence of points $\{y_0, y_1, y_2, y_3, \ldots\}$ on $\{t_0, t_0 + h, t_0 + 2h, t_0 + 3h, \ldots\}$ then

$$y_{n+1} = h \times f(t_0 + nh, y_n). \tag{5.4}$$

This sequence approaches the true solution y as $h \to 0$. Thus numerical methods, including the Runge-Kutta methods and the Euler method[2], step through this sequence $\{y_n\}$, choosing appropriate values of h and employing other methods of error reduction.

5.3 SOLVING WITH PYTHON

Here the `solve_ivp` method of the SciPy [80] library will be used to numerically solve the above models.

First the system of differential equations described in Equations 5.1, 5.2 and 5.3 must be defined. This is done using a Python function, where the first two arguments are the current time and the system state, respectively.

Python input

```
627  def derivatives(t, y, vaccine_rate, birth_rate=0.01):
628      """Defines the system of differential equations that describe
629      the epidemiology model.
630
631      Args:
632          t: a positive float
633          y: a tuple of three integers
634          vaccine_rate: a positive float <= 1
635          birth_rate: a positive float <= 1 (default: 0.01)
636
637      Returns:
638          A tuple containing dS, dI, and dR
639      """
640      infection_rate = 0.3
641      recovery_rate = 0.02
642      death_rate = 0.01
643      S, I, R = y
644      N = S + I + R
645      dSdt = (
646          -((infection_rate * S * I) / N)
647          + ((1 - vaccine_rate) * birth_rate * N)
648          - (death_rate * S)
```

---

[2]These methods are studied in the area of numerical analysis. A good textbook is [8].

```
649        )
650        dIdt = (
651            ((infection_rate * S * I) / N)
652            - (recovery_rate * I)
653            - (death_rate * I)
654        )
655        dRdt = (
656            (recovery_rate * I)
657            - (death_rate * R)
658            + (vaccine_rate * birth_rate * N)
659        )
660        return dSdt, dIdt, dRdt
```

Using this function returns the instantaneous rate of change for each of the three quantities, S, I and R. Starting at time 0.0, with 4 susceptible individuals, 1 infected individual, 0 recovered individuals, and a vaccine rate of 50%, gives:

Python input

```
661    print(derivatives(t=0.0, y=(4, 1, 0), vaccine_rate=0.5))
```

Python output

```
662    (-0.255, 0.21, 0.045)
```

this means that the number of susceptible individuals is expected to reduce by around 0.255 per time unit, the number of infected individuals to increase by 0.21 per time unit, and the number of recovered individuals to increase by 0.045 per time unit. After a tiny fraction of a time unit these quantities will change, and thus the rates of change will change.

The following function observes the system's behaviour over some time period, using SciPy's `solve_ivp` to numerically solve the system of differential equations:

Python input

```
663    from scipy.integrate import solve_ivp
664
665
666    def solve_ode(
667        derivative_function,
668        t_span,
```

```
669        y0=(2999, 1, 0),
670        vaccine_rate=0.85,
671        birth_rate=0.01,
672    ):
673        """Numerically solve the system of differential equations.
674
675        Args:
676            derivative_function: a function returning a tuple
677                                 of three floats
678            t_span: endpoints of the time range to integrate over
679            y0: a tuple of three integers (default: (2999, 1, 0))
680            vaccine_rate: a positive float <= 1 (default: 0.85)
681            birth_rate: a positive float <= 1 (default: 0.01)
682
683        Returns:
684            A tuple of four arrays
685        """
686        sol = solve_ivp(
687            derivative_function,
688            t_span,
689            y0,
690            args=(vaccine_rate, birth_rate),
691        )
692        t, S, I, R = sol.t, sol.y[0], sol.y[1], sol.y[2]
693        return t, S, I, R
```

This function can be used to investigate the difference in behaviour between a vaccination rate of 0% and a vaccination rate of 85%. The system will now be observed for two years, that is 730 days.

Begin with a vaccination rate of 0%:

Python input

```
694    t_span = [0, 730]
695    t, S, I, R = solve_ode(derivatives, t_span, vaccine_rate=0.0)
```

Now S, I and R are arrays of values of the stock levels of S, I and R over the time steps t. These can be plotted to visualise their behaviour, shown in Figure 5.2.

The number of infected individuals increases quickly, and in fact the rate of change increases as more individuals are infected. However this growth slows down as there are fewer susceptible individuals to infect. Due to the equal birth and death rates the overall population size remains constant; but after some time period (around 200 time units) the levels of susceptible, infected, and recovered individuals stabilise, and the disease becomes endemic. Once this occurs, around 10% of the population remain susceptible to the disease, 30% are infected, and 60% are recovered and immune.

Now with a vaccine rate of 85%:

```
Python input

696   t, S, I, R = solve_ode(derivatives, t_span, vaccine_rate=0.85)
```

The corresponding plot is shown in Figure 5.3.

With vaccination the disease remains endemic; however once steadiness occurs, around 10% of the population remain susceptible to the disease, 1.7% are infected and 88.3% are immune or recovered and immune.

This shows that vaccination lowers the percentage of the population living with the infection, which will lower the public healthcare system's costs. This saving will now be compared to the cost of providing the vaccination to the newborns.

The following function calculates the total cost to the public healthcare system, that is the sum of the medication costs for those living with the infection and the vaccination costs:

```
Python input

697   def daily_cost(
698       derivative_function=derivatives, vaccine_rate=0.85
699   ):
700       """Calculates the daily cost to the public health system after
701       2 years.
702
703       Args:
704           derivative_function: a function returning a tuple
705                                 of three floats
706           vaccine_rate: a positive float <= 1 (default: 0.85)
707
708       Returns:
709           the daily cost
710       """
711       birth_rate = 0.01
712       vaccine_cost = 220
713       medication_cost = 10
714       t_span = [0, 730]
715       t, S, I, R = solve_ode(
716           derivatives,
717           t_span,
```

```
718            vaccine_rate=vaccine_rate,
719            birth_rate=birth_rate,
720        )
721        N = S[-1] + I[-1] + R[-1]
722        daily_vaccine_cost = (
723            N * birth_rate * vaccine_rate * vaccine_cost
724        )
725        daily_meds_cost = I[-1] * medication_cost
726        return daily_vaccine_cost + daily_meds_cost
```

Now the total daily cost with and without vaccination can be compared. Without vaccinations:

Python input

```
727  cost = daily_cost(vaccine_rate=0.0)
728  print(round(cost, 2))
```

which gives

Python output

```
729  9002.33
```

Therefore without vaccinations, once the infection is endemic, the public health care system would expect to spend £9002.33 a day.

With a vaccination rate of 85%:

Python input

```
730  cost = daily_cost(vaccine_rate=0.85)
731  print(round(cost, 2))
```

which gives

Python output

```
732  6119.14
```

So vaccinating 85% of the population would cost the public health care system, once the infection is endemic £6119.14 a day. That is a saving of around 32%.

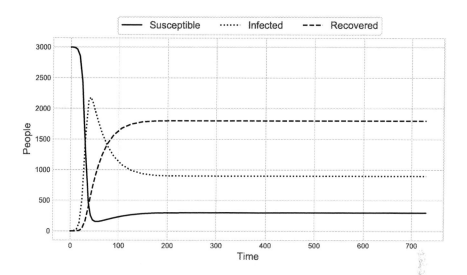

Figure 5.2 Stock levels without vaccination.

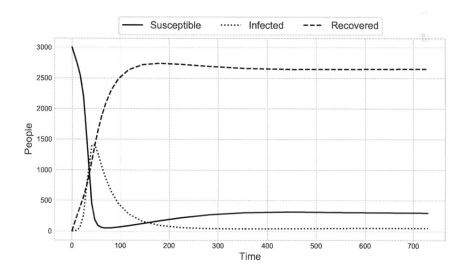

Figure 5.3 Stock levels with vaccination.

5.4   SOLVING WITH R

The deSolve [64] library will be used to numerically solve the above epidemiology models.

First the system of differential equations described in Equations 5.1, 5.2 and 5.3 must be defined. This is done using an R function, where the arguments are the current time, system state and a list of other parameters.

```
R input

733  #' Defines the system of differential equations that describe
734  #' the epidemiology model.
735  #'
736  #' @param t a positive float
737  #' @param y a tuple of three integers
738  #' @param parameters a vector with the vaccine_rate and birth_rate
739  #'
740  #' @return a list containing dS, dI, and dR
741  derivatives <- function(t, y, parameters){
742    infection_rate <- 0.3
743    recovery_rate <- 0.02
744    death_rate <- 0.01
745    with(
746      as.list(c(y, parameters)), {
747        N <- S + I + R
748        dSdt <- ( - ( (infection_rate * S * I) / N)
749          + ( (1 - vaccine_rate) * birth_rate * N)
750          - (death_rate * S))
751        dIdt <- ( ( (infection_rate * S * I) / N)
752          - (recovery_rate * I)
753          - (death_rate * I))
754        dRdt <- ( (recovery_rate * I)
755          - (death_rate * R)
756          + (vaccine_rate * birth_rate * N))
757        list(c(dSdt, dIdt, dRdt))
758      }
759    )
760  }
```

This function returns the instantaneous rate of change for each of the three quantities S, I and R. Starting at time 0.0, with 4 susceptible individuals, 1 infected individual, 0 recovered individuals, a vaccine rate of 50% and a birth rate of 0.01 gives:

```
761  ders <- derivatives(
762    t = 0,
763    y = c(S = 4, I = 1, R = 0),
764    parameters = c(vaccine_rate = 0.5, birth_rate = 0.01)
765  )
766  print(ders)
```

R output

```
767  [[1]]
768  [1] -0.255  0.210  0.045
```

The number of susceptible individuals is expected to reduce by around 0.255 per time unit, the number of infected individuals to increase by 0.21 per time unit and the number of recovered individuals to increase by 0.045 per time unit. After a tiny fraction of a time unit these quantities will change, and thus the rates of change will change.

The following function observes the system's behaviour over some time period, using the deSolve library to numerically solve the system of differential equations:

R input

```
769  library(deSolve)
770
771  #' Numerically solve the system of differential equations
772  #'
773  #' @param times an array of increasing positive floats
774  #' @param y0 list of integers (default: c(S=2999, I=1, R=0))
775  #' @param birth_rate a positive float <= 1 (default: 0.01)
776  #' @param vaccine_rate a positive float <= 1 (default: 0.85)
777  #'
778  #' @return a matrix of times, S, I and R values
779  solve_ode <- function(times,
780                        y0 = c(S = 2999, I = 1, R = 0),
781                        birth_rate = 0.01,
782                        vaccine_rate = 0.85){
783    params <- c(
784      birth_rate = birth_rate,
785      vaccine_rate = vaccine_rate
786    )
```

```
787    ode(
788        y = y0,
789        times = times,
790        func = derivatives,
791        parms = params
792    )
793  }
```

This function can be used to investigate the difference in behaviour between a vaccination rate of 0% and a vaccination rate of 85%. The system will be observed for two years, that is 730 days, in time steps of 0.01 days.

Begin with a vaccination rate of 0%:

R input

```
794  times <- seq(0, 730, by = 0.01)
795  out <- solve_ode(times, vaccine_rate = 0.0)
```

Now out, is a matrix with four columns, time, S, I and R, which are arrays of values of the time, points, and the stock levels of S, I and R over the time, respectively. These can be plotted to visualise their behaviour, shown in Figure 5.2[3].

The number of infected individuals increases quickly, and in fact the rate of change increases as more individuals are infected. However this growth slows down as there are fewer susceptible individuals to infect. Due to the equal birth and death rates the overall population size remains constant; but after some time period (around 200 time units) the levels of susceptible, infected, and recovered individuals stabilises, and the disease becomes endemic. Once this steadiness occurs, around 10% of the population remain susceptible to the disease, 30% are infected and 60% are recovered and immune.

Now with a vaccination rate of 85%:

R input

```
796  times <- seq(0, 730, by = 0.01)
797  out <- solve_ode(times, vaccine_rate = 0.85)
```

The corresponding plot is shown in Figure 5.3.

With vaccination, the disease remains endemic; however once steadiness occurs, around 10% of the population remain susceptible to the disease, 1.7% are infected and 88.3% are immune or recovered and immune.

---

[3]The particular figures shown in Figures 5.2 and 5.3 were created using Python and the Matplotlib library [30]. R has powerful plotting tools with packages such as ggplot2 [83].

This shows that vaccination lowers the percentage of the population living with the infection, which will lower the public healthcare system's costs. This saving will now be compared to the cost of providing the vaccination to the newborns.

The following function calculates the total cost to the public healthcare system, that is the sum of the medication costs for those living with the infection and the vaccination costs:

R input

```
798   #' Calculates the daily cost to the public health
799   #' system after 2 years
800   #'
801   #' @param derivative_function: a function returning a
802   #'                              list of three floats
803   #' @param vaccine_rate: a positive float <= 1 (default: 0.85)
804   #'
805   #' @return the daily cost
806   daily_cost <- function(derivative_function = derivatives,
807                          vaccine_rate = 0.85){
808     max_time <- 730
809     time_step <- 0.01
810     birth_rate <- 0.01
811     vaccine_cost <- 220
812     medication_cost <- 10
813     times <- seq(0, max_time, by = time_step)
814     out <- solve_ode(times, vaccine_rate = vaccine_rate)
815     N <- sum(tail(out[, c("S", "I", "R")], n = 1))
816     daily_vaccine_cost <- (
817       N * birth_rate * vaccine_rate * vaccine_cost
818     )
819     daily_medication_cost <- (
820       tail(out[, "I"], n = 1) * medication_cost
821     )
822     daily_vaccine_cost + daily_medication_cost
823   }
```

The total daily cost with and without vaccination will now be compared. Without vaccinations:

R input

```
824   cost <- daily_cost(vaccine_rate = 0.0)
825   print(cost)
```

which gives

```
R output
826   [1] 9000
```

Therefore without vaccinations, once the infection is endemic, the public health care system would expect to spend £9000 a day.

With a vaccination rate of 85%:

```
R input
827   cost <- daily_cost(vaccine_rate = 0.85)
828   print(cost)
```

which gives

```
R output
829   [1] 6119.034
```

So vaccinating 85% of newborns would cost the public healthcare system, once the infection is endemic £6119.04 a day. That is a saving of around 32%.

5.5 WIDER CONTEXT

System dynamics is an applied aspect of the more general mathematical field of dynamical systems. [55] gives a mathematical overview of the theory of dynamical systems and [8] is a good text on the numerical algorithms used to be able to observe the behaviour of these. For an overview of the type of application covered in this chapter, see [37].

In the field of operational research, Jay Forrester is recognised as the first person to use dynamical systems in the way shown in this chapter. His own account can be read in [22]. From Forrester's initial application to industry in 1961 [21] dynamical systems continue to be of use today in a wide range of areas. In [12] a case study of using dynamical systems for relevant modelling for the navy is described. [79] gives a literature review of the application area of healthcare. For example, [11] applies the same model from this chapter to the study of the COVID pandemic.

In order to be able to fully capture all the relevant details of the system to be modelled, an extension of system dynamics is to combine the methodology with discrete event simulation (see Chapter 3) in order to model discrete aspects of the system and/or agent-based modelling (see Chapter 7) in order to observe emergent or learned behaviours. This is known as hybrid simulation, and an overview is given in [7].

IV

Emergent Behaviour

CHAPTER 6

Game Theory

M OST of the time when modelling certain situations two approaches are valid: to make assumptions about the overall behaviour or to make assumptions about the detailed behaviour. The later can be thought of as measuring emergent behaviour. One tool used to do this is the study of interactive decision making: game theory.

6.1   PROBLEM

Consider a city council. Two electric taxi companies, company A and company B, are going to move in to the city and the city wants to ensure that the customers are best served by this new duopoly. The two taxi firms will be deciding how many vehicles to deploy: one, two or three. The city wants to encourage them to both use three as this ensures the highest level of availability to the population.

Some exploratory data analysis gives the following insights:

- if both companies use the same number of taxis, then they make the same profit which will go down slightly as the number of taxis goes up;

- if one company uses more taxis than the other then they make more profit.

The expected profits for the companies are given in Table 6.1.

Given these expected profits, the council wants to understand what is likely to happen and potentially give a financial incentive to each company to ensure their behaviour is in the population's interest. This would take the form of a fixed increase to the companies' profits, ϵ, to be found, if they put on three taxis, shown in Table 6.2

For example, from Table 6.2 it can be seen that if Company B chooses to use 3 vehicles while Company A chooses to only use 2 then Company B would get $\frac{17}{20} + \epsilon$ and Company A would get $\frac{1}{2}$ profits per hour. The question is: what value of ϵ ensures both companies use 3 vehicles?

6.2   THEORY

In the case of this city, the interaction can be modelled using a mathematical object called a normal form game, which here requires:

DOI: 10.1201/9780429328534-6

		Company B		
		1	2	3
Company A	1	1	$\frac{1}{2}$	$\frac{1}{3}$
	2	$\frac{3}{2}$	$\frac{19}{20}$	$\frac{1}{2}$
	3	$\frac{5}{3}$	$\frac{4}{5}$	$\frac{17}{20}$

		Company B		
		1	2	3
Company A	1	1	$\frac{3}{2}$	$\frac{5}{3}$
	2	$\frac{1}{2}$	$\frac{19}{20}$	$\frac{4}{5}$
	3	$\frac{1}{3}$	$\frac{1}{2}$	$\frac{17}{20}$

Table 6.1 Profits (in GBP per hour) of each taxi company based on the choice of vehicle number by all companies. The first table shows the profits for company A. The second table shows the profits for company B.

		Company B		
		1	2	3
Company A	1	1	$\frac{1}{2}$	$\frac{1}{3}$
	2	$\frac{3}{2}$	$\frac{19}{20}$	$\frac{1}{2}$
	3	$\frac{5}{3}+\epsilon$	$\frac{4}{5}+\epsilon$	$\frac{17}{20}+\epsilon$

		Company B		
		1	2	3
Company A	1	1	$\frac{3}{2}$	$\frac{5}{3}+\epsilon$
	2	$\frac{1}{2}$	$\frac{19}{20}$	$\frac{4}{5}+\epsilon$
	3	$\frac{1}{3}$	$\frac{1}{2}$	$\frac{17}{20}+\epsilon$

Table 6.2 Profits (in GBP per hour) of each taxi company based on the choice of vehicle number by all companies. The first table shows the profits for company A. The second table shows the profits for company B. The council's financial incentive ϵ is included.

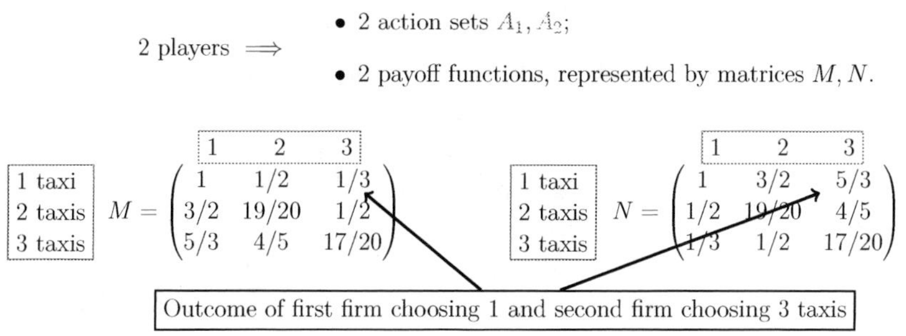

2 players $\implies$
- 2 action sets A_1, A_2;
- 2 payoff functions, represented by matrices M, N.

$$M = \begin{array}{c c} & \begin{array}{ccc} 1 & 2 & 3 \end{array} \\ \begin{array}{c} 1 \text{ taxi} \\ 2 \text{ taxis} \\ 3 \text{ taxis} \end{array} & \begin{pmatrix} 1 & 1/2 & 1/3 \\ 3/2 & 19/20 & 1/2 \\ 5/3 & 4/5 & 17/20 \end{pmatrix} \end{array} \qquad N = \begin{array}{c c} & \begin{array}{ccc} 1 & 2 & 3 \end{array} \\ \begin{array}{c} 1 \text{ taxi} \\ 2 \text{ taxis} \\ 3 \text{ taxis} \end{array} & \begin{pmatrix} 1 & 3/2 & 5/3 \\ 1/2 & 19/20 & 4/5 \\ 1/3 & 1/2 & 17/20 \end{pmatrix} \end{array}$$

Outcome of first firm choosing 1 and second firm choosing 3 taxis

Figure 6.1 Diagrammatic representation of the action sets and payoff matrices for the game.

1. a given collection of actors that make decisions (players);

2. options available to each player (actions);

3. a numerical value associated to each player for every possible choice of action made by all the players. This is the utility or reward.

This is called a normal form game and is formally defined by:

1. a finite set of N players;

2. action spaces for each player: $\{A_1, A_2, A_3, \ldots, A_N\}$;

3. utility functions that for each player $u_1, u_2, u_3, \ldots, u_N$ where $u_i : A_1 \times A_2 \times A_3 \ldots A_N \to \mathbb{R}$.

When $N = 2$ the utility function is often represented by a pair of matrices (1 for each player) with the same number of rows and columns. The rows correspond to the actions available to the first player and the columns to the actions available to the second player.

Given a pair of actions (a row and column), we can read the utilities to both players by looking at the corresponding entry of the corresponding matrix.

For this example, the two matrices would be:

$$M = \begin{pmatrix} 1 & 1/2 & 1/3 \\ 3/2 & 19/20 & 1/2 \\ 5/3 & 4/5 & 17/20 \end{pmatrix} \qquad N = M^T = \begin{pmatrix} 1 & 3/2 & 5/3 \\ 1/2 & 19/20 & 4/5 \\ 1/3 & 1/2 & 17/20 \end{pmatrix}$$

A diagram of the system is shown in Figure 6.1

A strategy corresponds to a way of choosing actions, this is represented by a probability vector. For the ith player, this vector v would be of size $|A_i|$ (the size of the action space) and v_i corresponds to the probability of choosing the ith action.

Both taxi firms always choosing to use 2 taxis (the second row/column) would correspond to both taxi firms choosing the strategy: $(0, 1, 0)$. If both companies use this strategy and the row player (who controls the rows) wants to improve their outcome, it is evident by inspecting the second column that the highest number is $19/20$: thus the row player has no reason to change what they are doing.

This is called a Nash equilibrium: when both players are playing a strategy that is the best response against the other.

An important fact is that a Nash equilibrium is guaranteed to exist. This was actually the theoretic result for which John Nash received a Nobel Prize[1] There are various algorithms that can be used for finding Nash equilibria, they involve a search through the pairs of spaces of possible strategies until pairs of best responses are found. Mathematical insight allows this do be done somewhat efficiently using algorithms that can be thought of as modifications of the algorithms used in linear programming (see Chapter 8). One such example is called the Lemke-Howson algorithm. A Nash equilibrium is not necessarily guaranteed to be arrived at through dynamic decision making. However, any stable behaviour that does emerge will be a Nash equilibrium, such emergent processes are the topics of evolutionary game theory[2], learning algorithms[3] and/or agent-based modelling, which will be covered in Chapter 7.

6.3 SOLVING WITH PYTHON

The first step is to write a function to create a game using the matrix of expected profits and any offset. The Nashpy [34] and Numpy [28] libraries will be used for this.

```
     Python input
830  import nashpy as nash
831  import numpy as np
832
833
834  def get_game(profits, offset=0):
835      """Return the game object with a given offset when 3 taxis are
836      provided.
837
838      Args:
839          profits: a matrix with expected profits
840          offset: a float (default: 0)
841
```

---

[1]John Nash proved the fact that any game has a Nash equilibrium in [50]. Discussions and proofs for particular cases can be found in good game theory text books such as [38].

[2]Evolutionary game theory was formalised in [63] although some of the work of Robert Axelrod is evolutionary in principle [1].

[3]An excellent text on learning in games is [23].

```
842        Returns:
843            A nashpy game object
844        """
845        new_profits = np.array(profits)
846        new_profits[2] += offset
847        return nash.Game(new_profits, new_profits.T)
```

This gives the game for the considered problem:

Python input

```
848   profits = np.array(
849       (
850           (1, 1 / 2, 1 / 3),
851           (3 / 2, 19 / 20, 1 / 2),
852           (5 / 3, 4 / 5, 17 / 20),
853       )
854   )
855   game = get_game(profits=profits)
856   print(game)
```

which gives:

Python output

```
857   Bi matrix game with payoff matrices:
858
859   Row player:
860   [[1.         0.5        0.33333333]
861    [1.5        0.95       0.5        ]
862    [1.66666667 0.8        0.85       ]]
863
864   Column player:
865   [[1.         1.5        1.66666667]
866    [0.5        0.95       0.8        ]
867    [0.33333333 0.5        0.85       ]]
```

The function `get_equilibria` which will directly compute the equilibria:

Python input

```
868  def get_equilibria(profits, offset=0):
869      """Return the equilibria for a given offset when 3 taxis are
870      provided.
871
872      Args:
873          profits: a matrix with expected profits
874          offset: a float (default: 0)
875
876      Returns:
877          A tuple of Nash equilibria
878      """
879      game = get_game(profits=profits, offset=offset)
880      return tuple(game.support_enumeration())
```

This can be used to obtain the equilibria in the game.

Python input

```
881  equilibria = get_equilibria(profits=profits)
```

The equilibria are:

Python input

```
882  for eq in equilibria:
883      print(eq)
```

giving:

Python output

```
884  (array([0., 1., 0.]), array([0., 1., 0.]))
885  (array([0., 0., 1.]), array([0., 0., 1.]))
886  (array([0. , 0.7, 0.3]), array([0. , 0.7, 0.3]))
```

There are 3 Nash equilibria: 3 possible pairs of behaviour that the 2 companies could stabilise at:

- the first equilibrium $((0, 1, 0), (0, 1, 0))$ corresponds to both firms always using 2 taxis;

- the second equilibrium $((0, 0, 1), (0, 0, 1))$ corresponds to both firms always using 3 taxis;

- the third equilibrium $((0, 0.7, 0.3), (0, 0.7, 0.3))$ corresponds to both firms using 2 taxis 70% of the time and 3 taxis otherwise.

A good thing to note is that the two taxi companies will never only provide a single taxi (which would be harmful to the customers).

The code below can be used to find the number of Nash equilibria for a given offset and stop when there is a single equilibrium:

Python input

```
887  offset = 0
888  while len(get_equilibria(profits=profits, offset=offset)) > 1:
889      offset += 0.01
```

This gives a final offset value of:

Python input

```
890  print(round(offset, 2))
```

Python output

```
891  0.15
```

and now confirm that the Nash equilibrium is where both taxi firms provide three vehicles:

Python input

```
892  print(get_equilibria(profits=profits, offset=offset))
```

giving:

Python output

```
893  ((array([0., 0., 1.]), array([0., 0., 1.])),)
```

Therefore, in order to ensure that the maximum amount of taxis are used, the council should offer a £0.15 per hour incentive to both taxi companies for putting on 3 taxis.

6.4 SOLVING WITH R

R does not have a uniquely appropriate library for the game considered here, we will choose to use Recon [9] which has functionality for finding the Nash equilibria for two player games when only considering pure strategies (where the players only choose to use a single action at a time).

```
R input
894   library(Recon)
895
896   #' Returns the equilibria in pure strategies
897   #' for a given offset
898   #'
899   #' @param profits: a matrix with expected profits
900   #' @param offset: a float (default: 0)
901   #'
902   #' @return a list of equilibria
903   get_equilibria <- function(profits, offset = 0){
904     new_profits <- rbind(
905       profits[c(1, 2), ],
906       profits[3, ] + offset
907     )
908     sim_nasheq(new_profits, t(new_profits))
909   }
```

This gives the pure Nash equilibria:

```
R input
910   profits <- rbind(
911     c(1, 1 / 2, 1 / 3),
912     c(3 / 2, 19 / 20, 1 / 2),
913     c(5 / 3, 4 / 5, 17 / 20)
914   )
915   eqs <- get_equilibria(profits = profits)
916   print(eqs)
```

which gives:

```
R output
917   $`Equilibrium 1`
918   [1] "2" "2"
919
920   $`Equilibrium 2`
921   [1] "3" "3"
```

There are 2 pure Nash equilibria: 2 possible pairs of behaviour that the two companies might converge to.

- the first equilibrium $((0,1,0),(0,1,0))$ corresponds to both firms always using 2 taxis;

- the second equilibrium $((0,0,1),(0,0,1))$ corresponds to both firms always using 3 taxis.

There is in fact a third Nash equilibrium where both taxi firms use 2 taxis 70% of the time and 3 taxis the rest of the time but Recon is unable to find Nash equilibria with mixed behaviour for games with more than two strategies.

As discussed, the council would like to offset the cost of 3 taxis so as to encourage the taxi company to provide a better service.

The code below gives the number of equilibria for a given offset and stops when there is a single equilibrium:

R input

```
922  offset <- 0
923  while (length(
924    get_equilibria(profits = profits, offset = offset)
925    ) > 1){
926    offset <- offset + 0.01
927  }
```

This gives a final offset value of:

R input

```
928  print(round(offset, 2))
```

R output

```
929  [1] 0.15
```

Now confirm that the Nash equilibrium is where both taxi firms provide three vehicles:

R input

```
930  eqs <- get_equilibria(profits = profits, offset = offset)
931  print(eqs)
```

giving:

```
R output

932   $`Equilibrium 1`
933   [1] "3" "3"
```

Therefore, in order to ensure that the maximum amount of taxis are used, the council should offer a £0.15 per hour incentive to both taxi companies for putting on 3 taxis.

6.5 WIDER CONTEXT

The definition of a normal form game here and the solution concept of Nash equilibrium are common starting points for the use of game theory as a study of emergent behaviour. Other topics include different forms of games, for example extensive form games, which are represented by trees and more explicitly model asynchronous decision making. Other solution concepts involve the specific study of the emergence mechanisms through models based on natural evolutionary process: Moran processes or replicator dynamics. A good text book to read on these topics is [38].

John Nash whose life was portrayed in the 2001 movie "A Beautiful Mind" (which is an adaptation of [48]) won the Nobel Prize for [50] in which he proved that a Nash equilibrium always exists. In [49] John Nash gives an application of game theory to a specific version of poker.

Another application of the concept of Nash equilibrium is [14] where the authors identify worst case scenarios for ambulance diversion: a practice where an emergency room will declare itself too full to accept new patients. When there are multiple emergency units serving a same population strategic behaviour becomes relevant. The authors of this paper identify the effect of this decentralised decision making and also propose an approach that is socially optimal: similarly to the taxi problem considered in this chapter.

A specific area of a lot of research in game theory is the study of cooperative behaviour. Axelrod [1] started this work with computer tournaments where he invited code submissions of strategies to play a game called the Iterated Prisoner's Dilemma. The outcome of this was an explanation for how cooperation can emerge in a complex system with a set of 5 rules, which included the need to be "nice". The conclusions of Axelrod's work and these set of rules continue to be examined and often refuted. For example, more recent work involving so-called Zero-Determinant strategies, which considered extortion as a mathematical concept [53] and also a possible outcome as opposed to cooperation. A review and systemic analysis of some of the research on behaviour, of which game theory is a subset, is given in [53].

CHAPTER 7

Agent-Based Simulation

S OMETIMES individual behaviours and interactions are well understood, and an understanding of how a whole population of such individuals might behave needed. For example, psychologists and economists may know a lot about how individual spenders and vendors behave in response to given stimuli, but an understanding of how these stimuli might effect the macro-economy is necessary. Agent-based simulation is a paradigm of thinking that allows such emergent population-level behaviour to be investigated from individual rules and interactions.

7.1 PROBLEM

Consider a city populated by two categories of household, for example a household might be fans of Cardiff City FC or Swansea City AFC[1]. Each household has a preference for living close to households of the same kind, and will move around the city while their preferences are not satisfied. How will these individual preferences affect the overall distribution of fans in the city?

7.2 THEORY

The problem considered here is considered a "classic" one for the paradigm of agent-based simulation, and is referred to as Schelling's segregation model. It features in Thomas Schelling's book "Micromotives and Macrobehaviours" [59], whose title neatly summarises the world view of agent-based modelling: we know, understand, determine, or can control individual micromotives; and from this we'd like to observe and understand macrobehaviours.

In general an agent-based model consists of two components, agents and an environment:

- agents are autonomous entities that will periodically choose to take one of a number of actions. These are chosen in order to maximise that agent's own given utility function;

- an environment contains a number of agents and defines how their interactions

---

[1]Swansea and Cardiff are two cities in South Wales with rival football clubs.

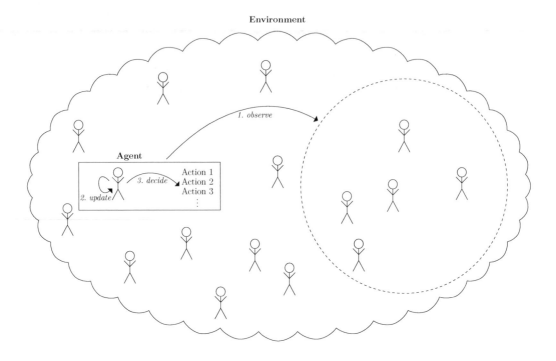

Figure 7.1 Representation of an agent interacting with its environment.

affect each other. The agents may be homogeneous or heterogeneous, and the relationships may change over time, possibly due to the actions taken by the agents.

In general, an agent will first observe a subset of its environment, for example it will consider some information about the agents it is currently close to. Then it will update some information about itself based on these observations. This could be recording relevant information from the observations, but could also include some learning, maybe considering its own previous actions. It will then decide on an action to take and carry out this action. This decision may be deterministic or random and/or based on its own attributes from some learning process; with the ultimate aim of maximising its own utility. In practice, a utility can be represented by a function that maps the environment to some numeric value. This process happens to all agents in the environment, possibly simultaneously. This is summarised in Figure 7.1.

For the football team supporters problem, each household is an agent. The environment is the city. Each household's utility function is to satisfy their preference of living next to at least a given number of households supporting the same team as them. Their choices of action are to move house or not to move house.

As a simplification the city will be modelled as a 50×50 grid. Each cell of the grid is a house that can either contain a household of Cardiff City FC supporters, or contain a household of Swansea City AFC supporters. A house's neighbours are assumed to be the houses adjacent to it, horizontally, vertically and diagonally. For mathematical simplicity, it is also assumed that the grid is a torus, where houses in

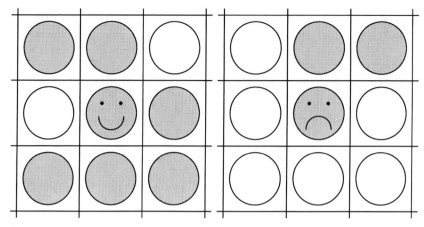

(a) A happy household, with 6 similar neighbours ($\frac{6}{8} > p = 0.5$)

(b) An unhappy household, with 2 similar neighbours ($\frac{2}{8} < p = 0.5$)

Figure 7.2 Example of a household happy and unhappy with its neighbours, when $p = 0.5$. Households supporting Cardiff City FC are shaded grey, households supporting Swansea City AFC are white.

the top row are vertically adjacent to the bottom row and houses in the rightmost column are horizontally adjacent to the leftmost column.

Every household has a preference p. This corresponds to the minimum proportion of neighbours they are happy to live next to. Figure 7.2 shows a household of Cardiff City FC supporters that are happy with their neighbours, and not happy with their neighbours, when $p = 0.5$. Households supporting Cardiff City FC are shaded grey.

The original problem stated that households move around the city whenever they are unhappy with their neighbours. This long process of selling, searching for and buying houses can be simplified to randomly pairing two unhappy households and swapping their houses. In fact, this can be simplified to consider the houses themselves as agents, who swap households with each other.

Therefore the model logic is:

1. initialise the model: fill each house in the grid with either a household of Cardiff City FC or Swansea City AFC supporters with probability 0.5 each;

2. at each discrete time step, for every house:

   (a) consider their household's neighbours (*observe*);
   (b) determine if the household is happy (*update*);
   (c) if unhappy (*decide*), swap household with another randomly chosen house with an unhappy household (*action*).

After a number of time steps the overall structure of the city can be observed. This agent-based model only explicitly defines individual behaviours and interactions; therefore any population-level behaviour that may have emerged would occur without explicit definition.

7.3 SOLVING WITH PYTHON

Agent-based modelling lends itself well to a programming paradigm called object-orientated programming. This paradigm lets a number of *objects* from a set of instructions called a *class* to be built. These objects can both store information (in Python these are called *attributes*), and do things (in Python these are called *methods*). Object-orientated programming allows for the creation of new classes, which can be used to implement the individual behaviours of an agent-based model.

For this problem two classes will be built: a House and a City for them to live in. The following libraries will be used:

```
Python input
934  import random
935  import itertools
936  import numpy as np
```

Now to define the City:

```
Python input
937  class City:
938      def __init__(self, size, threshold):
939          """Initialises the City object.
940
941          Args:
942              size: an integer number of rows and columns
943              threshold: float between 0 and 1 representing the
944                  minimum acceptable proportion of similar neighbours
945          """
946          self.size = size
947          sides = range(size)
948          self.coords = itertools.product(sides, sides)
949          self.houses = {
950              (x, y): House(x, y, threshold, self)
951              for x, y in self.coords
952          }
953
954      def run(self, n_steps):
955          """Runs the simulation of a number of time steps.
956
957          Args:
958              n_steps: an integer number of steps
```

```
959              """
960          for turn in range(n_steps):
961              self.take_turn()
962
963      def take_turn(self):
964          """Swaps all sad households."""
965          sad = [h for h in self.houses.values() if h.sad()]
966          random.shuffle(sad)
967          i = 0
968          while i <= len(sad) / 2:
969              sad[i].swap(sad[-i])
970              i += 1
971
972      def mean_satisfaction(self):
973          """Finds the average household satisfaction.
974
975          Returns:
976              The average city's household satisfaction
977          """
978          return np.mean(
979              [h.satisfaction() for h in self.houses.values()]
980          )
```

This defines a class, a template or a set of instructions that can be used to create instances, called objects. For the considered problem only one instance of the City class will be needed. However, it is useful to be able to produce more in order to run multiple trials with different random seeds. This class contains four methods: __init__, run, take_turn and mean_satisfaction.

The __init__ method is run whenever the object is first created and initialises the object. In this case, it sets a number of attributes.

- First the square grid's size is defined, which is the number of rows and columns of houses it contains;

- next the coords are defined, a list of tuples representing all the possible coordinates of the grid, this uses the itertools library for efficient iteration;

- finally houses is defined, a dictionary with grid coordinates as keys, and instances of the House class.

The run method runs the simulation. For each n_steps number of discrete time steps, the city runs the method take_turn. In this method, first a list of all the houses is created with households that are unhappy with their neighbours; these are put in a random order using the random library; and then working inwards from the boundary houses with sad households are paired up and swap households.

The last method defined here is the `mean_satisfaction` method, which is only used to observe any emergent behaviour. This calculates the average satisfaction of all the houses in the grid, using the numpy library for convenience.

In order to be able to create an instance of the above class, we need to define a House class:

```
Python input
```

```python
981  class House:
982      def __init__(self, x, y, threshold, city):
983          """Initialises the House object.
984
985          Args:
986              x: the integer x-coordinate
987              y: the integer y-coordinate
988              threshold: a number between 0 and 1 representing
989                  the minimum acceptable proportion of similar
990                  neighbours
991              city: an instance of the City class
992          """
993          self.x = x
994          self.y = y
995          self.threshold = threshold
996          self.kind = random.choice(["Cardiff", "Swansea"])
997          self.city = city
998
999      def satisfaction(self):
1000         """Determines the household's satisfaction level.
1001
1002         Returns:
1003             A proportion
1004         """
1005         same = 0
1006         for x, y in itertools.product([-1, 0, 1], [-1, 0, 1]):
1007             ax = (self.x + x) % self.city.size
1008             ay = (self.y + y) % self.city.size
1009             same += self.city.houses[ax, ay].kind == self.kind
1010         return (same - 1) / 8
1011
1012     def sad(self):
1013         """Determines if the household is sad.
1014
1015         Returns:
1016             a Boolean
```

```
1017            """
1018            return self.satisfaction() < self.threshold
1019
1020        def swap(self, house):
1021            """Swaps two households.
1022
1023            Args:
1024                house: the house object to swap household with
1025            """
1026            self.kind, house.kind = house.kind, self.kind
```

It contains four methods: __init__, satisfaction, sad and swap.

The __init__ methods sets a number of attributes at the time the object is created: the house's x and y coordinates (its column and row numbers on the grid); its threshold which corresponds to p; its kind which is randomly chosen between having a Cardiff City FC supporting household and a Swansea City AFC supporting household; and finally its city, an instance of the City class, shared by all the houses.

The satisfaction method loops through each of the house's neighbouring cells in the city grid, counts the number of neighbours that are of the same kind as itself and returns this as a proportion. Then the sad method returns a boolean indicating if the household's satisfaction is below the minimum threshold.

Finally the swap method takes another house object and swaps their household kinds.

A function to create and run one of these simulations will now be written with a given random seed, threshold and number of steps. This function returns the resulting mean happiness:

Python input

```
1027   def find_mean_happiness(seed, size, threshold, n_steps):
1028       """Create and run an instance of the simulation.
1029
1030       Args:
1031           seed: the random seed to use
1032           size: an integer number of rows and columns
1033           threshold: a number between 0 and 1 representing
1034               the minimum acceptable proportion of similar
1035               neighbours
1036           n_steps: an integer number of steps
1037
1038       Returns:
1039           The average city's household satisfaction after
```

```
1040            n_steps
1041        """
1042        random.seed(seed)
1043        C = City(size, threshold)
1044        C.run(n_steps)
1045        return C.mean_satisfaction()
```

Now consider each household with a threshold of 0.65, and compare the mean happiness after 0 steps and 100 steps. First 0 steps:

Python input

```
1046    initial_happiness = find_mean_happiness(
1047        seed=0, size=50, threshold=0.65, n_steps=0
1048    )
1049    print(initial_happiness)
```

Python output

```
1050    0.4998
```

This is well below the minimum threshold of 0.65, and so on average households are unhappy. After 100 steps:

Python input

```
1051    final_happiness = find_mean_happiness(
1052        seed=0, size=50, threshold=0.65, n_steps=100
1053    )
1054    print(final_happiness)
```

Python output

```
1055    0.9078
```

After 100 time steps the average satisfaction level is much higher. In fact, it is much higher than each individual household's threshold. Now consider that this satisfaction level is really a level of how similar each households' neighbours are; it is actually a level of segregation. This was the central premise of Schelling's original model [59] that overall emergent segregation levels are much higher than any individuals' personal preference for segregation.

(a) At the beginning.

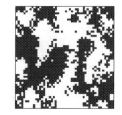

(b) After 20 time steps.

(c) After 100 time steps.

Figure 7.3 Plotted results from the Python code.

More analysis methods can be added, including plotting functions. Figure 7.3 shows the grid at the beginning, after 20 time steps and after 100 time steps, with households supporting Cardiff City FC in grey, and those supporting Swansea City AFC in white. It visually shows the households segregating over time.

7.4 SOLVING WITH R

Agent-based modelling lends itself well to a programming paradigm called object-orientated programming. This paradigm lets a number of *objects* from a set of instructions called a *class* to be built. These objects can both store information (in the R package used here these are called *fields*) and do things (in the R package used here these are called *methods*). Object-orientated programming allows for the creation of new classes, which can be used to implement the individual behaviours of an agent-based model.

There are a number of ways of doing object-orientated programming in R. In this chapter, a package called R6 will be used here.

For this problem two classes will be built: a House and a City for them to live in. Now to define the City[2]

```
R input

1056  library(R6)
1057  City <- R6Class("City", list(
1058    size = NA,
1059    houses = NA,
1060    initialize = function(size, threshold) {
1061      self$size <- size
1062      self$houses <- c()
1063      for (x in 1:size) {
```

---

[2]Note that no documentation is included in the definition of the class and the style is inconsistent with the other chapters in this book. The authors do not know of any widely accepted convention for documenting classes in R, furthermore the text of this chapter serves as detailed documentation in this context.

```
1064        row <- c()
1065        for (y in 1:size) {
1066          row <- c(row, House$new(x, y, threshold, self))
1067        }
1068        self$houses <- rbind(self$houses, row)
1069      }
1070    },
1071    run = function(n_steps) {
1072      if (n_steps > 0) {
1073        for (turn in 1:n_steps) {
1074          self$take_turn()
1075        }
1076      }
1077    },
1078    take_turn = function() {
1079      sad <- c()
1080      for (house in self$houses) {
1081        if (house$sad()) {
1082          sad <- c(sad, house)
1083        }
1084      }
1085      sad <- sample(sad)
1086      num_sad <- length(sad)
1087      i <- 1
1088      while (i <= num_sad / 2) {
1089        sad[[i]]$swap(sad[[num_sad - i]])
1090        i <- i + 1
1091      }
1092    },
1093    mean_satisfaction = function() {
1094      mean(sapply(self$houses, function(x) x$satisfaction()))
1095    }
1096  ) )
```

This defines an R6 class, a template or a set of instructions that can be used to create instances of it, called objects. For this model only one instance of the City class is needed, although it may be useful to be able to produce more in order to run multiple trials with different random seeds. This class contains four methods: initialize, run, take_turn and mean_satisfaction.

The initialize method is run at the time the object is first created. It initialises the object by setting a number of its fields:

- First the square grid's size is defined, which is the number of rows and columns of houses it contains;

- then the **houses** are defined by iteratively repeating the **rbind** function to create a two-dimensional vector of instances of the, yet to be defined, **House** class, representing the houses themselves.

The **run** method runs the simulation. For each discrete time step from 1 to **n_steps**, the world runs the method **take_turn**. In this method, a list of all the houses with households that are unhappy with their neighbours is created; these are put in a random order and then working inwards from the boundary, houses with sad households are paired up and swap households.

The last method defined here is the **mean_satisfaction** method, which is used to observe the emergent behaviour. This calculates the average satisfaction of all the houses in the grid.

In order to be able to create an instance of the above class, a **House** class is needed:

R input

```
1097  House <- R6Class("House", list(
1098    x = NA,
1099    y = NA,
1100    threshold = NA,
1101    city = NA,
1102    kind = NA,
1103    initialize = function(x = NA,
1104                          y = NA,
1105                          threshold = NA,
1106                          city = NA) {
1107      self$x <- x
1108      self$y <- y
1109      self$threshold <- threshold
1110      self$city <- city
1111      self$kind <- sample(c("Cardiff", "Swansea"), 1)
1112    },
1113    satisfaction = function() {
1114      same <- 0
1115      for (x in -1:1) {
1116        for (y in -1:1) {
1117          ax <- ( (self$x + x - 1) %% self$city$size) + 1
1118          ay <- ( (self$y + y - 1) %% self$city$size) + 1
1119          if (self$city$houses[[ax, ay]]$kind == self$kind) {
1120            same <- same + 1
1121          }
1122        }
1123      }
1124      (same - 1) / 8
```

```
1125      },
1126      sad = function() {
1127        self$satisfaction() < self$threshold
1128      },
1129      swap = function(house) {
1130        old <- self$kind
1131        self$kind <- house$kind
1132        house$kind <- old
1133      }
1134    ) )
```

It contains four methods: `initialize`, `satisfaction`, `sad` and `swap`.

The `initialize` methods sets a number of the class' fields when the object is created: the house's x and y coordinates (its column and row numbers on the grid); its `threshold` which corresponds to p; its `kind` which is randomly chosen between having a Cardiff City FC supporting household and a Swansea City AFC supporting household; and finally its `city`, an instance of the `City` class, shared by all the houses.

The `satisfaction` method loops through each of the house's neighbouring cells in the city grid, counts the number of neighbours that are of the same kind as itself, and returns this as a proportion. The `sad` method returns a boolean indicating of the household's satisfaction is below its minimum threshold.

Finally the `swap` method takes another house object and swaps their household kinds.

A function to create and run one of these simulations will now be written with a given random seed, threshold, and number of steps. This function return the resulting mean happiness:

R input

```
1135    #' Create and run an instance of the simulation.
1136    #'
1137    #' @param seed: the random seed to use
1138    #' @param size: an integer number of rows and columns
1139    #' @param threshold: a number between 0 and 1 representing
1140    #'    the minimum acceptable proportion of similar neighbours
1141    #' @param n_steps: an integer number of steps
1142    #'
1143    #' @return The average city's household satisfaction
1144    #'    after n_steps
1145    find_mean_happiness <- function(seed,
1146                                    size,
1147                                    threshold,
```

```
1148                                    n_steps){
1149     set.seed(seed)
1150     city <- City$new(size, threshold)
1151     city$run(n_steps)
1152     city$mean_satisfaction()
1153   }
```

Now consider each household with a threshold of 0.65, and compare the mean happiness after 0 steps and 100 steps. First 0 steps:

R input

```
1154   initial_happiness <- find_mean_happiness(
1155     seed = 0,
1156     size = 50,
1157     threshold = 0.65,
1158     n_steps = 0
1159   )
1160   print(initial_happiness)
```

R output

```
1161   [1] 0.4956
```

This is well below the minimum threshold of 0.65, and so on average households are unhappy here. Let's run the simulation for 100 generations and see how this changes:

R input

```
1162   final_happiness <- find_mean_happiness(
1163     seed = 0,
1164     size = 50,
1165     threshold = 0.65,
1166     n_steps = 100
1167   )
1168   print(final_happiness)
```

R output

```
1169   [1] 0.9338
```

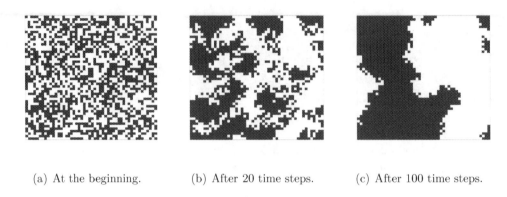

(a) At the beginning.   (b) After 20 time steps.   (c) After 100 time steps.

Figure 7.4   Plotted results from the R code.

After 100 time steps the average satisfaction has increased. It is now actually much higher than each individual household's threshold. This satisfaction level can be considered as a level of how similar each households' neighbours are, and so it is actually a level of segregation. This was the central premise of Schelling's original model [59], that overall emergent segregation levels are much higher than any individuals' personal preference for segregation.

More analysis methods can be added, including plotting functions. Figure 7.4 shows the grid at the beginning, after 20 time steps, and after 100 time steps, with households supporting Cardiff City FC in grey, and those supporting Swansea City AFC in white. It shows the households segregating over time.

7.5   WIDER CONTEXT

The simulations described in this chapter come under the larger umbrella term of multi agent systems, which discusses the theory of systems with multiple independent agents interacting with one another. A good source on the topic is [62].

The model described here is called Schelling's segregation model, and was first described in [59]. Another model considered as classic in this domain is a model of a flock of birds presented in [54], otherwise referred to as Boids, where the behaviours of flocks of birds are understood by capturing the individual interactions between individual birds. Conway's Game of Life, described in [24] is another classic, which comes under the banner of cellular automata. Here cells on a grid either become alive or dead depending on a certain simplistic set of rules. Emergent behaviours observed due to these rules include self replicating as well as oscillating structures. In the 1970s agent-based tournaments were held by Robert Axelrod [1], which was the first of a number of studies using agent-based modelling and game theory (see Chapter 6) to understand the emergence of cooperative behaviours.

In recent years, similar methodologies have been used in a variety of applications such as [15] which models parents' choice of school, in [57] archaeological population migration and trade dynamics are modelled, and [31] offers a systematic literature review for the use of agent-based modelling of autonomous vehicles.

V

Optimisation

CHAPTER 8

Linear Programming

F INDING the best configuration of some system can be challenging, especially when there is a seemingly endless amount of possible solutions. Optimisation techniques are a way to mathematically derive solutions that maximise or minimise some objective function, subject to a number of feasibility constraints. When all components of the problem can be written in a linear way, then linear programming is one technique that can be used to find the solution.

8.1 PROBLEM

A university runs 14 modules over three subjects: Art, Biology and Chemistry. Each subject runs core modules and optional modules. Table 8.1 gives the module numbers for each of these.

The university is required to schedule examinations for each of these modules. The university would like the exams to be scheduled using the least amount of time slots. However not all modules can be scheduled at the same time as they share some students:

- All art modules share students;

- all biology modules share students;

Art Core	Biology Core	Chemistry Core
M00	M05	M09
M01	M06	M10
Art Optional	**Biology Optional**	**Chemistry Optional**
M02	M07	M11
M03	M08	M12
M04		M13

Table 8.1 List of modules on offer at the university.

DOI: 10.1201/9780429328534-8

- all chemistry modules share students;

- biology students have the option of taking optional modules from chemistry, so all biology modules may share students with optional chemistry modules;

- chemistry students have the option of taking optional modules from biology, so all chemistry modules may share students with optional biology modules;

- biology students have the option of taking core art modules, and so all biology modules may share students with core art modules.

How can every exam be scheduled with no clashes, using the least amount of time slots?

8.2 THEORY

Linear programming is a method that solves a type of optimisation problem of a number of variables by making use of some concepts of higher dimensional geometry [10]. Optimisation here refers to finding the variable that gives either the maximum or minimum of some linear function, called the objective function.

Linear programming employs algorithms such as the Simplex method to efficiently search some feasible convex region, stopping at the optimum. To do this, an objective function and constraints need to be defined.

To illustrate this a classic two-dimensional example will be used: £50 of profit can be made on each tonne of paint A produced and £60 profit on each tonne of paint B produced. A tonne of paint A needs 4 tonnes of component X and 5 tonnes of component Y. A tonne of paint B needs 6 tonnes of component X and 4 tonnes of component Y. Only 24 tonnes of X and 20 tonnes of Y are available per day. How much of paint A and paint B should be produced to maximise profit?

This is formulated as a linear objective function, representing total profit, that is to be maximised; and two linear constraints, representing the availability of components X and Y. They are written as:

$$\text{Maximise: } 50A + 60B \tag{8.1}$$
$$\text{Subject to:}$$
$$4A + 6B \leq 24 \tag{8.2}$$
$$5A + 4B \leq 20 \tag{8.3}$$

Now this is a linear system in two-dimensional space with coordinates A and B. These are called the decision variables, what is required are the values of A and B that optimise the objective function given by expression 8.1.

Inequalities 8.2 and 8.3 correspond to the amount of component X and Y available per day. These, along with the additional constraints that a negative amount of paint cannot be produced ($A \geq 0$ and $B \geq 0$), form a convex region, shown in Figure 8.1. This shaded region shows the pairs of values of A and B which are feasible, that is they satisfy the constraints.

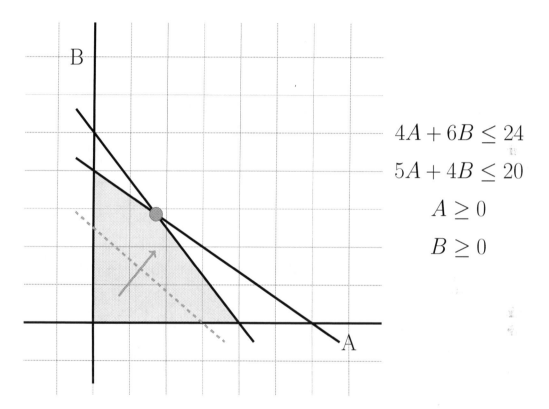

$$4A + 6B \leq 24$$
$$5A + 4B \leq 20$$
$$A \geq 0$$
$$B \geq 0$$

Figure 8.1 Visual representation of the paint linear programming problem. The feasible convex region is shaded in grey; the objective function with arbitrary value is shown in a dashed line.

Expression 8.1 corresponds to the total profit, which is the value to be maximised. As a line in two-dimensional space, this expression fixes its gradient, but its value determines the size of the y-intercept. Therefore optimising this function corresponds to pushing a line with that gradient to its furthest extreme within the feasible region, demonstrated in Figure 8.1. Therefore for this problem the optimum occurs in a particular vertex of the feasible region, at $A = \frac{12}{7}$ and $B = \frac{20}{7}$.

This works well as A and B can take any real value in the feasible region. Some problems must be formulated as integer linear programming problems where the decision variables are restricted to integers. There are a number of methods that can help adapt a real solution to an integer solution. These include cutting planes, which introduce new constraints around the real solution to force an integer value; and branch and bound methods, where we iteratively convert decision variables to their closest two integers and remove any infeasible solutions [10].

Both Python and R have libraries that carry out linear and integer programming algorithms. When solving these kinds of problems, formulating them as linear systems is the most important challenge.

Consider again the exam scheduling problem from Section 8.1, which will now be formulated as an integer linear programming problem. Define M as the set of all modules to be scheduled and define T as the set of possible time slots. At worst each exam is scheduled for a different day; thus $|T| = |M| = 14$ in this case. Let $\{X_{mt}$ for $m \in M$ and $t \in T\}$ be a set of binary decision variables, that is $X_{mt} = 1$ if module m is scheduled for time t, and 0 otherwise.

There are six distinct sets of modules in which exams cannot be scheduled simultaneously: A_c, A_o representing core and optional art modules respectively B_c, B_o representing core and optional biology modules respectively and C_c, C_o representing core and optional chemistry modules respectively. Therefore $M = A_c \cup A_o \cup B_c \cup B_o \cup C_c \cup C_o$.

Additionally there are further clashes between these sets:

- No modules in $A_c \cup A_o$ can be scheduled together as they may share students; this is given by the constraint in inequality 8.7.

- No modules in $B_c \cup B_o \cup A_c$ can be scheduled together as they may share students, given by inequality 8.8.

- No modules in $B_c \cup B_o \cup C_o$ can be scheduled together as they may share students, given by inequality 8.9.

- No modules in $B_o \cup C_c \cup C_o$ can be scheduled together as they may share students, given by inequality 8.10.

Define $\{Y_t$ for $t \in T\}$ as a set of auxiliary binary decision variables, where Y_t is 1 if time slot t is being used. This is enforced by Inequality 8.5.

Equation 8.6 ensures all modules are scheduled once and once only. Thus altogether the formulation becomes:

$$\text{Minimise: } \sum_{t \in T} Y_j \tag{8.4}$$

Subject to:

$$\frac{1}{|M|} \sum_{m \in M} X_{mt} \leq Y_j \text{ for all } j \in T \tag{8.5}$$

$$\sum_{t \in T} X_{mt} = 1 \text{ for all } m \in M \tag{8.6}$$

$$\sum_{m \in A_c \cup A_o} X_{mt} \leq 1 \text{ for all } t \in T \tag{8.7}$$

$$\sum_{m \in B_c \cup B_o \cup A_c} X_{mt} \leq 1 \text{ for all } t \in T \tag{8.8}$$

$$\sum_{m \in B_c \cup B_o \cup C_o} X_{mt} \leq 1 \text{ for all } t \in T \tag{8.9}$$

$$\sum_{m \in B_o \cup C_c \cup C_o} X_{mt} \leq 1 \text{ for all } t \in T \tag{8.10}$$

Another common way to define this linear programming problem is by representing the coefficients of the constraints as a matrix. That is:

$$\text{Minimise: } c^T Z \tag{8.11}$$

Subject to:

$$AZ \star b \tag{8.12}$$

where Z is a vector representing the decision variables, c is the coefficients of Z in the objective function, A is the matrix of the coefficients of Z in the constraints, b is the vector of the right-hand side of the constraints and $\star$ represents either $\leq$, $=$ or $\geq$ as required.

As Z is a one-dimensional vector of decision variables, the matrix X and the vector Y can be "flattened" together to form this new variable. This is done by first ordering X then Y, within that ordering by time slot, then within that ordering by module number. Therefore:

$$Z_{|M|t+m} = X_{mt} \tag{8.13}$$

$$Z_{|M|^2+m} = Y_m \tag{8.14}$$

where t and m are indices starting at 0. For example, Z_{17} would correspond to $X_{3,2}$, the decision variable representing whether module number 4 is scheduled on day 3; Z_{208} would correspond to Y_{12}, the decision variable representing whether there is an exam scheduled for day 12.

Parameters c, A and b can be determined by using this same conversion from the model in Equations 8.4 to 8.10. The vector c would be $|M|^2$ zeroes followed by $|M|$ ones. The vector b would be zeroes for all the rows representing Equation 8.5, and ones for all other constraints.

8.3   SOLVING WITH PYTHON

In this book the Python library Pulp [44] will be used to formulate and solve the integer programming problem. First a function to create the binary problem variables for a given set of times and modules is needed:

```python
1170  import pulp
1171
1172
1173  def get_variables(modules, times):
1174      """Returns the binary variables for a given timetabling
1175      problem.
1176
1177      Args:
1178          modules: The complete collection of modules to be
1179                   timetabled.
1180          times: The collection of available time slots.
1181
1182      Returns:
1183          A tuple containing the decision variables x and y.
1184      """
1185      xshape = (modules, times)
1186      x = pulp.LpVariable.dicts("X", xshape, cat=pulp.LpBinary)
1187      y = pulp.LpVariable.dicts("Y", times, cat=pulp.LpBinary)
1188      return x, y
```

The specific modules and times relating to the problem can now be used to obtain the corresponding variables:

```python
1189  Ac = [0, 1]
1190  Ao = [2, 3, 4]
1191  Bc = [5, 6]
1192  Bo = [7, 8]
1193  Cc = [9, 10]
1194  Co = [11, 12, 13]
1195  modules = Ac + Ao + Bc + Bo + Cc + Co
1196  times = range(14)
1197  x, y = get_variables(modules=modules, times=times)
```

Now y is a dictionary of binary decision variables, with keys as elements of the list times. Y_3 corresponds to the third day:

Python input

```
1198   print(y[3])
```

Python output

```
1199   Y_3
```

While x is a dictionary of dictionaries of binary decision variables, with keys as elements of the lists modules and times. $X_{2,5}$ is the variable corresponding to module 2 being scheduled on day 5:

Python input

```
1200   print(x[2][5])
```

Python output

```
1201   X_2_5
```

The next step is to create a Pulp object with the corresponding variables, objective function and constraints and solve it. This is done with the following function:

Python input

```
1202   def get_solution(Ac, Ao, Bc, Bo, Cc, Co, times):
1203       """Returns the binary variables corresponding to the solution
1204       of given timetabling problem.
1205
1206       Args:
1207           Ac: The set of core art modules
1208           Ao: The set of optional art modules
1209           Bc: The set of core biology modules
1210           Bo: The set of optional biology modules
1211           Cc: The set of core chemistry modules
1212           Co: The set of optional chemistry modules
1213           times: The collection of available time slots.
1214
1215       Returns:
1216           A tuple containing the decision variables x and y.
1217       """
```

```
1218    modules = Ac + Ao + Bc + Bo + Cc + Co
1219    x, y = get_variables(modules=modules, times=times)
1220    prob = pulp.LpProblem("ExamScheduling", pulp.LpMinimize)
1221    objective_function = sum([y[day] for day in times])
1222    prob += objective_function
1223
1224    M = 1 / len(modules)
1225    for day in times:
1226        prob += M * sum(x[m][day] for m in modules) <= y[day]
1227        prob += sum([x[mod][day] for mod in Ac + Ao]) <= 1
1228        prob += sum([x[mod][day] for mod in Bc + Bo + Co]) <= 1
1229        prob += sum([x[mod][day] for mod in Bc + Bo + Ac]) <= 1
1230        prob += sum([x[mod][day] for mod in Cc + Co + Bo]) <= 1
1231
1232    for mod in modules:
1233        prob += sum(x[mod][day] for day in times) == 1
1234
1235    prob.solve()
1236    return x, y
```

Using this, the solution x of the original problem can be obtained:

Python input

```
1237    x, y = get_solution(
1238        Ac=Ac, Ao=Ao, Bc=Bc, Bo=Bo, Cc=Cc, Co=Co, times=times
1239    )
```

These can be inspected, for example x_{25} is a boolean variable relating to if module 2 is scheduled on the 5th day.

Python input

```
1240    print(x[2][5].value())
```

Python output

```
1241    0.0
```

This was assigned the value 0, and so module 2 was not scheduled for that day. However, module 2 was scheduled for day 9:

Python input

```
1242    print(x[2][9].value())
```

Python output

```
1243    1.0
```

This was assigned a value of 1, and so module 2 was scheduled for that day.
The following function creates a readable schedule:

Python input

```
1244    def get_schedule(x, y, Ac, Ao, Bc, Bo, Cc, Co, times):
1245        """Returns a human readable schedule corresponding to the
1246        solution of given timetabling problem.
1247
1248        Args:
1249            Ac: The set of core art modules
1250            Ao: The set of optional art modules
1251            Bc: The set of core biology modules
1252            Bo: The set of optional biology modules
1253            Cc: The set of core chemistry modules
1254            Co: The set of optional chemistry modules
1255            times: The collection of available time slots.
1256
1257        Returns:
1258            A string with the schedule
1259        """
1260        modules = Ac + Ao + Bc + Bo + Cc + Co
1261
1262        schedule = ""
1263        for day in times:
1264            if y[day].value() == 1:
1265                schedule += f"\nDay {day}: "
1266                for mod in modules:
1267                    if x[mod][day].value() == 1:
1268                        schedule += f"{mod}, "
1269        return schedule
```

Thus:

```
     Python input
1270 schedule = get_schedule(
1271     x=x,
1272     y=y,
1273     times=times,
1274     Ac=Ac,
1275     Ao=Ao,
1276     Bc=Bc,
1277     Bo=Bo,
1278     Cc=Cc,
1279     Co=Co,
1280 )
1281 print(schedule)
```

gives:

```
     Python output
1282 Day 0: 1, 12,
1283 Day 5: 0, 13,
1284 Day 6: 11,
1285 Day 7: 4, 6, 10,
1286 Day 8: 3, 5, 9,
1287 Day 9: 2, 7,
1288 Day 13: 8,
```

The order of the days does not matter here, but 7 days are required in order to schedule all exams with no clashes, with at most three exams scheduled each day.

8.4 SOLVING WITH R

The R package ROI, the R Optimization Infrastructure will be used here. This is a library of code that acts as a front end to a number of other solvers that need to be installed externally, allowing a range of optimisation problems to be solved with a number of different solvers. The solver that will be used here is called the CBC [20] MILP Solver, which needs to be installed. The rcbc [61] package is also necessary but cannot be installed in the usual way. Installation instructions for both depend on the operating system and can be found at the documentation page for the ROI [74] package[1].

The ROI package requires that the linear programming problem is represented in its matrix form, with a one-dimensional array of decision variables. Therefore the

---

[1]As of the time of writing, this can be found at https://roi.r-forge.r-project.org/installation.html

form of the model described at the end of Section 8.2 will be used. Functions that define the coefficients of the objective function c, the coefficient matrix A, the vector of the right-hand side of the constraints b and the vector of equality or inequalities directions $\star$ are needed.

First the objective function:

```
R input

1289   #' Writes the row of coefficients for the objective function
1290   #'
1291   #' @param n_modules: the number of modules to schedule
1292   #' @param n_days: the maximum number of days to schedule
1293   #'
1294   #' @return the objective function row to minimise
1295   write_objective <- function(n_modules, n_days){
1296      all_days <- rep(0, n_modules * n_days)
1297      Ys <- rep(1, n_days)
1298      append(all_days, Ys)
1299   }
```

For 3 modules and 3 days:

```
R input

1300   objectives <- write_objective(n_modules = 3, n_days = 3)
1301   print(objectives)
```

Which gives the following array, corresponding to the coefficients of the array Z for Equation 8.4.

```
R output

1302   [1] 0 0 0 0 0 0 0 0 0 1 1 1
```

The following function is used to write one row of that coefficients matrix, for a given day, for a given set of clashes, corresponding to Inequalities 8.7 to 8.10:

```
      R input

1303  #' Writes the constraint row dealing with clashes
1304  #'
1305  #' @param clashes: a vector of module indices that all cannot
1306  #'                  be scheduled at the same time
1307  #' @param day: an integer representing the day
1308  #' @param n_days the maximum number of days to schedule
1309  #' @param n_modules the number of modules to schedule
1310  #' @return the constraint row corresponding to that set of
1311  #'                clashes on that day
1312  write_X_clashes <- function(clashes, day, n_days, n_modules){
1313    today <- rep(0, n_modules)
1314    today[clashes] = 1
1315    before_today <- rep(0, n_modules * (day - 1))
1316    after_today <- rep(0, n_modules * (n_days - day))
1317    all_days <- c(before_today, today, after_today)
1318    full_coeffs <- c(all_days, rep(0, n_days))
1319    full_coeffs
1320  }
```

where `clashes` is an array containing the module numbers of a set of modules that may all share students.

The following function is used to write one row of the coefficients matrix, for each module, ensuring that each module is scheduled on one day and one day only, corresponding to Equation 8.6:

```
      R input

1321  #' Writes the constraint row to ensure that every module is
1322  #' scheduled once and only once
1323  #'
1324  #' @param module: an integer representing the module
1325  #' @param n_days the maximum number of days to schedule
1326  #' @param n_modules the number of modules to schedule
1327  #' @return the constraint row corresponding to scheduling a
1328  #'                module on only one day
1329  write_X_requirements <- function(module, n_days, n_modules){
1330    today <- rep(0, n_modules)
1331    today[module] = 1
1332    all_days <- rep(today, n_days)
1333    full_coeffs <- c(all_days, rep(0, n_days))
1334    full_coeffs
1335  }
```

The following function is used to write one row of the coefficients matrix corresponding to the auxiliary constraints of Inequality 8.5:

```
R input
1336  #' Writes the constraint row representing the Y variable,
1337  #' whether at least one exam is scheduled on that day
1338  #'
1339  #' @param day: an integer representing the day
1340  #' @param n_days the maximum number of days to schedule
1341  #' @param n_modules the number of modules to schedule
1342  #' @return the constraint row corresponding to creating Y
1343  write_Y_constraints <- function(day, n_days, n_modules){
1344      today <- rep(1, n_modules)
1345      before_today <- rep(0, n_modules * (day - 1))
1346      after_today <- rep(0, n_modules * (n_days - day))
1347      all_days <- c(before_today, today, after_today)
1348      all_Ys <- rep(0, n_days)
1349      all_Ys[day] = -n_modules
1350      full_coeffs <- append(all_days, all_Ys)
1351      full_coeffs
1352  }
```

Finally the following function uses all previous functions to assemble a coefficient matrix. It loops through the parameters for each constraint row required, uses the appropriate function to create the row of the coefficient matrix and sets the appropriate inequality direction ($\leq, =, \geq$) and the value of the right-hand side. It returns all three components:

```
R input
1353  #' Writes all the constraints as a matrix, column of
1354  #' inequalities, and right hand side column.
1355  #'
1356  #' @param list_clashes: a list of vectors with sets of modules
1357  #         that cannot be scheduled at the same time
1358  #' @param n_days the maximum number of days to schedule
1359  #' @param n_modules the number of modules to schedule
1360  #' @return f.con the LHS of the constraints as a matrix
1361  #' @return f.dir the directions of the inequalities
1362  #' @return f.rhs the values of the RHS of the inequalities
1363  write_constraints <- function(list_clashes, n_days, n_modules){
1364      all_rows <- c()
1365      all_dirs <- c()
1366      all_rhss <- c()
1367      n_rows <- 0
```

```
1368
1369    for (clash in list_clashes){
1370      for (day in 1:n_days){
1371        clashes <- write_X_clashes(clash, day, n_days, n_modules)
1372        all_rows <- append(all_rows, clashes)
1373        all_dirs <- append(all_dirs, "<=")
1374        all_rhss <- append(all_rhss, 1)
1375        n_rows <- n_rows + 1
1376      }
1377    }
1378    for (module in 1:n_modules){
1379      reqs <- write_X_requirements(module, n_days, n_modules)
1380      all_rows <- append(all_rows, reqs)
1381      all_dirs <- append(all_dirs, "==")
1382      all_rhss <- append(all_rhss, 1)
1383      n_rows <- n_rows + 1
1384    }
1385    for (day in 1:n_days){
1386      Yconstraints <- write_Y_constraints(day, n_days, n_modules)
1387      all_rows <- append(all_rows, Yconstraints)
1388      all_dirs <- append(all_dirs, "<=")
1389      all_rhss <- append(all_rhss, 0)
1390      n_rows <- n_rows + 1
1391    }
1392    f.con <- matrix(all_rows, nrow = n_rows, byrow = TRUE)
1393    f.dir <- all_dirs
1394    f.rhs <- all_rhss
1395    list(f.con, f.dir, f.rhs)
1396  }
```

For demonstration, with 2 modules and 2 possible days, with the single constraint that both modules cannot be scheduled at the same time, then:

R input

```
1397  write_constraints(
1398    list_clashes = list(c(1, 2)),
1399    n_days = 2,
1400    n_modules = 2
1401  )
```

This would give 3 components:

- a coefficient matrix of the left-hand side of the constraints, A (rows 1 and 2 corresponding to the clash on days 1 and 2, row 3 ensuring module 1 is scheduled on one day only, row 4 ensuring module 2 is scheduled on one day only, and rows 5 and 6 defining the decision variables Y);

- an array of direction of the constraint inequalities, $\star$;

- and an array of the right-hand side values of the constraints, b.

```
R output
[[1]]
     [,1] [,2] [,3] [,4] [,5] [,6]
[1,]   1    1    0    0    0    0
[2,]   0    0    1    1    0    0
[3,]   1    0    1    0    0    0
[4,]   0    1    0    1    0    0
[5,]   1    1    0    0   -2    0
[6,]   0    0    1    1    0   -2

[[2]]
[1] "<=" "<=" "==" "==" "<=" "<="

[[3]]
[1] 1 1 1 1 0 0
```

Now, the problem will be solved. First some parameters, including the sets of modules that all share students, that is the list of clashes are needed:

```
R input
n_modules = 14
n_days = 14
Ac <- c(0, 1)
Ao <- c(2, 3, 4)
Bc <- c(5, 6)
Bo <- c(7, 8)
Cc <- c(9, 10)
Co <- c(11, 12, 13)
list_clashes <- list(
    c(Ac, Ao),
    c(Bc, Bo, Co),
    c(Bc, Bo, Ac),
    c(Bo, Cc, Co)
)
```

Then, the functions defined above are used to create the objective function and the 3 elements of the constraints:

```
R input
1430  constraints <- write_constraints(
1431    list_clashes = list_clashes,
1432    n_days = n_days,
1433    n_modules = n_modules
1434  )
1435  f.con <- constraints[[1]]
1436  f.dir <- constraints[[2]]
1437  f.rhs <- constraints[[3]]
1438  f.obj <- write_objective(n_modules = n_modules, n_days = n_days)
```

Finally, once these objects are in place, the ROI library is used to construct an optimisation problem object:

```
R input
1439  library(ROI)
1440
1441  milp <- OP(
1442    objective = L_objective(f.obj),
1443    constraints = L_constraint(L = f.con, dir = f.dir, rhs = f.rhs),
1444    types = rep("B", length(f.obj)),
1445    maximum = FALSE
1446  )
```

This creates an OP object from our objective row f.obj, and our constraints, which are made up from the three components f.con, f.dir and f.rhs. When creating this object the types as binary variables are indicated (an array of "B" for each decision variable). The objective function is to be minimised so maximum = FALSE is used.

Now to solve:

```
R input
1447  sol <- ROI_solve(milp)
```

The solver will output information about the solve process and runtime.

```
R input
1448  print(sol$solution)
```

R output

```
1449    [1]  0 0 0 0 0 0 0 0 0 0 0 0 0 0 0 0 0 0 1 0 0 0 0 0 0 1 0 0 0 0
1450   [30]  0 0 0 0 0 0 0 0 0 0 0 0 0 0 0 0 0 0 0 0 0 0 0 0 0 0 0 0 0
1451   [59]  0 0 0 0 0 0 0 0 0 0 0 1 0 0 0 0 0 0 0 0 0 0 0 1 0 0 0 0 0
1452   [88]  0 0 0 0 0 0 0 0 0 0 0 0 0 0 0 0 0 1 0 0 0 0 0 0 0 0 0 0 0
1453  [117]  0 0 0 0 0 0 0 0 0 0 0 0 0 0 0 0 0 1 0 0 0 0 0 0 0 0 0 1 0 0
1454  [146]  0 0 0 0 0 0 0 1 0 0 1 0 0 0 1 0 0 1 0 0 1 0 0 0 0 1 0 0 0 0 0
1455  [175]  0 0 0 0 0 0 0 0 0 0 0 0 1 0 0 0 0 1 0 0 0 0 0 1 0 0 0 1 0
1456  [204]  1 0 1 1 1 0 1
```

This gives the values of each of the Z decision variables. We know the structure of this, that is the first 14 variables are the modules scheduled for day 1, and so on. The following code prints a readable schedule:

R input

```
1457  #' Gives a human readable schedule corresponding to the
1458  #' solution of a given timetable problem.
1459  #'
1460  #' @param sol: a solution to the timetabling problem
1461  #' @param n_modules: the number of modules to schedule
1462  #' @param n_days: the maximum number of days to schedule
1463  #'
1464  #' @return A string with the schedule
1465  get_schedule <- function(sol, n_days, n_modules){
1466    schedule = ""
1467    for (day in 1:n_days){
1468      if (sol$solution[(n_days * n_modules) + day] == 1){
1469        schedule <- paste(schedule, "\n", "Day", day, ":")
1470        for (module in 1:n_modules){
1471          var <- ((day - 1) * n_modules) + module
1472          if (sol$solution[var] == 1){
1473            schedule <- paste(schedule, module)
1474          }
1475        }
1476      }
1477    }
1478    schedule
1479  }
```

Thus:

```
     ┌─────────────────────────────────────────────────────────┐
     │  R input                                                │
     ├─────────────────────────────────────────────────────────┤
1480 │ schedule <- get_schedule(                               │
1481 │   sol = sol,                                            │
1482 │   n_days = n_days,                                      │
1483 │   n_modules = n_modules                                 │
1484 │ )                                                       │
1485 │ cat(schedule)                                           │
     └─────────────────────────────────────────────────────────┘
```

gives:

```
     ┌─────────────────────────────────────────────────────────┐
     │  R output                                               │
     ├─────────────────────────────────────────────────────────┤
1486 │ "Day 2 : 4 11"                                          │
1487 │ "Day 6 : 1 12"                                          │
1488 │ "Day 8 : 7"                                             │
1489 │ "Day 10 : 8"                                            │
1490 │ "Day 11 : 3 13"                                         │
1491 │ "Day 12 : 2 6 9 14"                                     │
1492 │ "Day 14 : 5 10"                                         │
     └─────────────────────────────────────────────────────────┘
```

This gives that 7 days are the minimum required to schedule the 14 exams without clashes, with either 1, 2 or 4 exams scheduled on each day.

8.5 WIDER CONTEXT

The overview given here on linear programming covers a wide breath of the subject although not much depth. For specific algorithmic approaches to the underlying algorithms and problem types, such as branch and bound and cutting plane methods as well as some minor extensions, see [10, 68].

The efficiency of linear programming as well as the ability to model linear situations implies that it is often used for a variety of applications. Theatre scheduling as one such application is given in [27]. Scheduling is indeed a wide-ranging subapplication of linear programming which can also be used to schedule sport seasons [17].

Other applications include the transportation problem [16], which can be used to find a best allocation of a fleet of delivery vehicles; the fire station location problem [60] used to minimise travel times to at-risk areas; and the bin packing problem [29] in which a number of, possibly irregular, shapes are packed into the smallest possible number of bins.

CHAPTER 9

Heuristics

I T is often necessary to find the most desirable choice from a large, or indeed, infinite set of options. Sometimes this can be done using exact techniques but often this is not possible and finding an almost perfect choice quickly is just as good. This is where the field of heuristics comes in to play.

9.1 PROBLEM

A delivery company needs to deliver goods to 13 different stops. They need to find a route for a driver that stops at each of the stops once only, then returns to the first stop, the depot.

The stops are drawn in Figure 9.1.

The relevant information is the pairwise distances between each of the stops, which is given by the distance matrix in equation (9.1).

$$
d = \begin{bmatrix}
0 & 35 & 35 & 29 & 70 & 35 & 42 & 27 & 24 & 44 & 58 & 71 & 69 \\
35 & 0 & 67 & 32 & 72 & 40 & 71 & 56 & 36 & 11 & 66 & 70 & 37 \\
35 & 67 & 0 & 63 & 64 & 68 & 11 & 12 & 56 & 77 & 48 & 67 & 94 \\
29 & 32 & 63 & 0 & 93 & 8 & 71 & 56 & 8 & 33 & 84 & 93 & 69 \\
70 & 72 & 64 & 93 & 0 & 101 & 56 & 56 & 92 & 81 & 16 & 5 & 69 \\
35 & 40 & 68 & 8 & 101 & 0 & 76 & 62 & 11 & 39 & 91 & 101 & 76 \\
42 & 71 & 11 & 71 & 56 & 76 & 0 & 15 & 65 & 81 & 40 & 60 & 94 \\
27 & 56 & 12 & 56 & 56 & 62 & 15 & 0 & 50 & 66 & 41 & 58 & 82 \\
24 & 36 & 56 & 8 & 92 & 11 & 65 & 50 & 0 & 39 & 81 & 91 & 74 \\
44 & 11 & 77 & 33 & 81 & 39 & 81 & 66 & 39 & 0 & 77 & 79 & 37 \\
58 & 66 & 48 & 84 & 16 & 91 & 40 & 41 & 81 & 77 & 0 & 20 & 73 \\
71 & 70 & 67 & 93 & 5 & 101 & 60 & 58 & 91 & 79 & 20 & 0 & 65 \\
69 & 37 & 94 & 69 & 69 & 76 & 94 & 82 & 74 & 37 & 73 & 65 & 0
\end{bmatrix}
\tag{9.1}
$$

The value d_{ij} gives the travel distance between stops i and j. For example, $d_{23} = 67$ indicates that the distance between the 2nd and 3rd stop in the route is 67.

The delivery company would like to find the route around the 13 stops that gives the smallest overall travel distance.

9.2 THEORY

This problem is called a travelling salesman problem, which can often be inefficient to solve using exact methods [43]. Heuristics are a family of methods that can be used to find a *sufficiently good* solution, though not necessarily the optimal solution, where the emphasis is on prioritising computational efficiency.

DOI: 10.1201/9780429328534-9

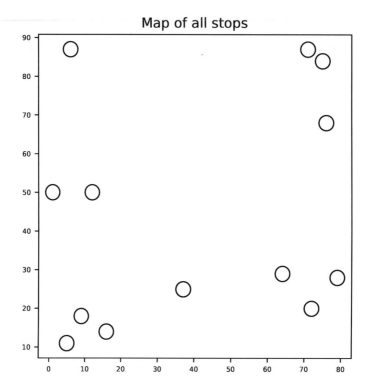

Figure 9.1 The positions of the required stops.

The heuristic approach taken here will be to use a neighbourhood search algorithm. This algorithm works by considering a given potential solution, evaluating it and then trying another potential solution *close* to it. What *close* means depends on different approaches and problems: it is referred to as the neighbourhood. When a new solution is considered *good*[1] then the search continues from the neighbourhood of this new solution.

For this problem, the steps are to first represent a possible solution, that is a given route between all the potential stops as a *tour*. If there are 3 total stops the tour must start and stop at the first one then there are two possible tours:

$$t \in \{(1, 2, 3, 1), (1, 3, 2, 1)\}$$

Given a distance matrix d such that d_{ij} is the distance between stop i and j the total cost of a tour is given by:

$$C(t) = \sum_{i=1}^{n} d_{t_i, t_{i+1}}$$

Thus, with:

$$d = \begin{pmatrix} 0 & 1 & 3 \\ 1 & 0 & 15 \\ 3 & 3 & 7 \end{pmatrix}$$

We have:

$$\begin{aligned} C((1, 2, 3, 1)) &= d_{12} + d_{23} + d_{31} = 1 + 15 + 3 = 19 \\ C((1, 3, 2, 1)) &= d_{13} + d_{32} + d_{21} = 3 + 3 + 1 = 7 \end{aligned}$$

In general, the neighbourhood search can be written down as:

1. Start with a given tour: t.

2. Evaluate $C(t)$.

3. Identify a new $\tilde{t}$ from t and accept it as a replacement for t if $C(\tilde{t}) < C(t)$.

4. Repeat the 3rd step until some stopping condition is met.

This is shown diagrammatically in Figure 9.2.

A number of stopping conditions can be used including some specific overall cost or a number of total iterations of the algorithm.

The neighbourhood of a tour t is taken as some set of tours that can be obtained from t using a specific and computationally efficient **neighbourhood operator**. To illustrate two such neighbourhood operators, consider the following tour on 7 stops:

---

[1]'Good' is again a term that depends on the approach and problem.

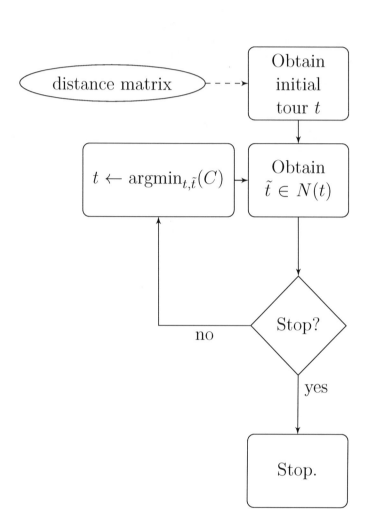

Figure 9.2 The general neighbourhood search algorithm. $N(t)$ refers to some neighbourhood of t.

$$t = (0, 1, 2, 3, 4, 5, 6, 0)$$

One possible neighbourhood is to choose 2 stops at random and swap. For example, the tour $\tilde{t}^{(1)} \in N(t)$ is obtained by swapping the stop labelled 2 and the stop labelled 5.

$$\tilde{t}^{(1)} = (0, 1, 5, 3, 4, 2, 6, 0)$$

Another possible neighbourhood is to choose 2 stops at random and reverse the order of all stops between (including) those two stops. For example, the tour $\tilde{t}^{(2)} \in N(t)$ is obtained by reversing the order of all stops between the stop labelled 2 and the stop labelled 5.

$$\tilde{t}^{(2)} = (0, 1, 5, 4, 3, 2, 6, 0)$$

Examples of these tours are shown in Figure 9.3.

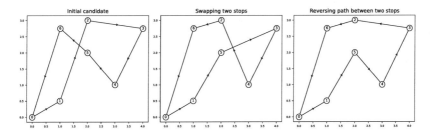

Figure 9.3 The effect of two neighbourhood operators on t. $\tilde{t}^{(1)}$ is obtained by swapping stops labelled 2 and 5. $\tilde{t}^{(2)}$ is obtained by reversing the path between stops labelled 2 and 5.

9.3 SOLVING WITH PYTHON

To solve this problem using Python, functions will be written that match the first three steps in Section 9.2. The first step is to write the `get_initial_candidate` function that creates an initial tour.

Python input

```
1493  import numpy as np
1494
1495
1496  def get_initial_candidate(number_of_stops, seed):
1497      """Return an random initial tour.
1498
1499      Args:
```

```
1500        number_of_stops: The number of stops
1501        seed: An integer seed.
1502
1503    Returns:
1504        A tour starting an ending at stop with index 0.
1505    """
1506    internal_stops = list(range(1, number_of_stops))
1507    np.random.seed(seed)
1508    np.random.shuffle(internal_stops)
1509    return [0] + internal_stops + [0]
```

This gives a random tour on 13 stops:

Python input

```
1510    number_of_stops = 13
1511    seed = 0
1512    initial_candidate = get_initial_candidate(
1513        number_of_stops=number_of_stops,
1514        seed=seed,
1515    )
1516    print(initial_candidate)
```

Python output

```
1517    [0, 7, 12, 5, 11, 3, 9, 2, 8, 10, 4, 1, 6, 0]
```

To be able to evaluate any given tour its cost must be found. Here `get_cost` does this:

Python input

```
1518    def get_cost(tour, distance_matrix):
1519        """Return the cost of a tour.
1520
1521        Args:
1522            tour: A given tuple of successive stops.
1523            distance_matrix: The distance matrix of the problem.
1524
1525        Returns:
1526            The cost
1527        """
```

```
1528    return sum(
1529        distance_matrix[current_stop, next_stop]
1530        for current_stop, next_stop in zip(tour[:-1], tour[1:])
1531    )
```

Python input

```
1532  distance_matrix = np.array(
1533      (
1534          (0, 35, 35, 29, 70, 35, 42, 27, 24, 44, 58, 71, 69),
1535          (35, 0, 67, 32, 72, 40, 71, 56, 36, 11, 66, 70, 37),
1536          (35, 67, 0, 63, 64, 68, 11, 12, 56, 77, 48, 67, 94),
1537          (29, 32, 63, 0, 93, 8, 71, 56, 8, 33, 84, 93, 69),
1538          (70, 72, 64, 93, 0, 101, 56, 56, 92, 81, 16, 5, 69),
1539          (35, 40, 68, 8, 101, 0, 76, 62, 11, 39, 91, 101, 76),
1540          (42, 71, 11, 71, 56, 76, 0, 15, 65, 81, 40, 60, 94),
1541          (27, 56, 12, 56, 56, 62, 15, 0, 50, 66, 41, 58, 82),
1542          (24, 36, 56, 8, 92, 11, 65, 50, 0, 39, 81, 91, 74),
1543          (44, 11, 77, 33, 81, 39, 81, 66, 39, 0, 77, 79, 37),
1544          (58, 66, 48, 84, 16, 91, 40, 41, 81, 77, 0, 20, 73),
1545          (71, 70, 67, 93, 5, 101, 60, 58, 91, 79, 20, 0, 65),
1546          (69, 37, 94, 69, 69, 76, 94, 82, 74, 37, 73, 65, 0),
1547      )
1548  )
1549  cost = get_cost(
1550      tour=initial_candidate,
1551      distance_matrix=distance_matrix,
1552  )
1553  print(cost)
```

Python output

```
1554  827
```

Now a function for neighbourhood operator will be written, swap_stops, that swaps two stops in a given tour.

Python input

```python
1555  def swap_stops(tour):
1556      """Return a new tour by swapping two stops.
1557
1558      Args:
1559          tour: A given tuple of successive stops.
1560
1561      Returns:
1562          A tour
1563      """
1564      number_of_stops = len(tour) - 1
1565      i, j = np.random.choice(range(1, number_of_stops), 2)
1566      new_tour = list(tour)
1567      new_tour[i], new_tour[j] = tour[j], tour[i]
1568      return new_tour
```

Applying this neighbourhood operator to the initial candidate gives:

Python input

```python
1569  print(swap_stops(initial_candidate))
```

which swaps the 10th and 12th stops:

Python output

```python
1570  [0, 7, 12, 5, 11, 3, 9, 2, 8, 1, 4, 10, 6, 0]
```

Now all the tools are in place to build a tool to carry out the neighbourhood search `run_neighbourhood_search`.

Python input

```python
1571  def run_neighbourhood_search(
1572      distance_matrix,
1573      iterations,
1574      seed,
1575      neighbourhood_operator=swap_stops,
1576  ):
1577      """Returns a tour by carrying out a neighbourhood search.
1578
```

```
1579        Args:
1580            distance_matrix: the distance matrix
1581            iterations: the number of iterations for which to
1582                        run the algorithm
1583            seed: a random seed
1584            neighbourhood_operator: the neighbourhood operator
1585                                    (default: swap_stops)
1586
1587        Returns:
1588            A tour
1589        """
1590        number_of_stops = len(distance_matrix)
1591        candidate = get_initial_candidate(
1592            number_of_stops=number_of_stops,
1593            seed=seed,
1594        )
1595        best_cost = get_cost(
1596            tour=candidate,
1597            distance_matrix=distance_matrix,
1598        )
1599        for _ in range(iterations):
1600            new_candidate = neighbourhood_operator(candidate)
1601            cost = get_cost(
1602                tour=new_candidate,
1603                distance_matrix=distance_matrix,
1604            )
1605            if cost <= best_cost:
1606                best_cost = cost
1607                candidate = new_candidate
1608
1609        return candidate
```

Now running this for 1000 iterations:

```
Python input

1610  number_of_iterations = 1000
1611
1612  solution_with_swap_stops = run_neighbourhood_search(
1613      distance_matrix=distance_matrix,
1614      iterations=number_of_iterations,
1615      seed=seed,
1616      neighbourhood_operator=swap_stops,
```

```
1617  )
1618  print(solution_with_swap_stops)
```

gives:

Python output

```
1619  [0, 7, 2, 8, 5, 3, 1, 9, 12, 11, 4, 10, 6, 0]
```

This has a cost:

Python input

```
1620  cost = get_cost(
1621      tour=solution_with_swap_stops,
1622      distance_matrix=distance_matrix,
1623  )
1624  print(cost)
```

Python output

```
1625  362
```

Therefore, using this particular algorithm, a pretty good route is found, with a total distance of 362.

It is important to note that this may not be the optimal route, and different algorithms may produce better solutions. For example, one way to modify the algorithm is to use a different neighbourhood operator. Instead of swapping two stops, reverse the path between those two stops. This corresponds to an algorithm called the "2-opt" algorithm[2]. The reverse_path function does this:

Python input

```
1626  def reverse_path(tour):
1627      """Return a new tour by reversing the path between two stops.
1628
1629      Args:
1630          tour: A given tuple of successive stops.
1631
1632      Returns:
```

---

[2]The 2-opt algorithm was first published in [13].

```
1633            A tour
1634        """
1635        number_of_stops = len(tour) - 1
1636        stops = np.random.choice(range(1, number_of_stops), 2)
1637        i, j = sorted(stops)
1638        new_tour = tour[:i] + tour[i : j + 1][::-1] + tour[j + 1 :]
1639        return new_tour
```

Applying this neighbourhood operator to the initial candidate gives:

Python input

```
1640  print(reverse_path(initial_candidate))
```

which reverses the order between the 3rd and the 11th stop:

Python output

```
1641  [0, 7, 4, 10, 8, 2, 9, 3, 11, 5, 12, 1, 6, 0]
```

Now running the neighbourhood search for 1000 iterations using the reverse_path neighbourhood operator:

Python input

```
1642  solution_with_reverse_path = run_neighbourhood_search(
1643      distance_matrix=distance_matrix,
1644      iterations=number_of_iterations,
1645      seed=seed,
1646      neighbourhood_operator=reverse_path,
1647  )
1648  print(solution_with_reverse_path)
```

gives:

Python output

```
1649  [0, 8, 5, 3, 1, 9, 12, 11, 4, 10, 6, 2, 7, 0]
```

This now gives a different route. Importantly, the costs differ substantially:

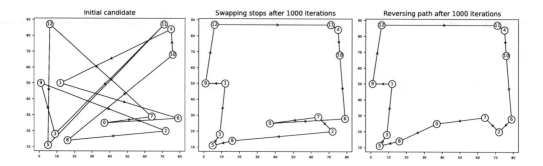

Figure 9.4 The final tours obtained by using the neighbourhood search in Python.

```
1650  cost = get_cost(
1651      tour=solution_with_reverse_path,
1652      distance_matrix=distance_matrix,
1653  )
1654  print(cost)
```

which gives:

```
1655  299
```

This improves on the solution found using the `swap_stops` operator. Figure 9.4 shows the final obtained routes given by both approaches.

9.4 SOLVING WITH R

To solve this problem using R, functions will be written that match the first three steps in Section 9.2.

The first step is to write the `get_initial_candidate` function that creates an initial tour:

```
R input
```

```
1656   #' Return an random initial tour.
1657   #'
1658   #' @param number_of_stops The number of stops.
1659   #' @param seed An integer seed.
1660   #'
1661   #' @return A tour starting an ending at stop with index 0.
1662   get_initial_candidate <- function(number_of_stops, seed){
1663     internal_stops <- 1:(number_of_stops - 1)
1664     set.seed(seed)
1665     internal_stops <- sample(internal_stops)
1666     c(0, internal_stops, 0)
1667   }
```

This gives a random tour on 13 stops:

```
R input
```

```
1668   number_of_stops <- 13
1669   seed <- 1
1670   initial_candidate <- get_initial_candidate(
1671     number_of_stops = number_of_stops,
1672     seed = seed)
1673   print(initial_candidate)
```

```
R output
```

```
1674   [1]  0  9  4  7  1  2  5  3  8  6 11 12 10  0
```

To be able to evaluate any given tour its cost must be found. Here `get_cost` does this:

```
R input
```

```
1675   #' Return the cost of a tour
1676   #'
1677   #' @param tour A given vector of successive stops.
1678   #' @param seed The distance matrix of the problem.
1679   #'
1680   #' @return The cost
1681   get_cost <- function(tour, distance_matrix){
```

```
1682    pairs <- cbind(tour[-length(tour)], tour[-1]) + 1
1683    sum(distance_matrix[pairs])
1684  }
```

R input

```
1685  distance_matrix <- rbind(
1686         c(0, 35, 35, 29, 70, 35, 42, 27, 24, 44, 58, 71, 69),
1687         c(35, 0, 67, 32, 72, 40, 71, 56, 36, 11, 66, 70, 37),
1688         c(35, 67, 0, 63, 64, 68, 11, 12, 56, 77, 48, 67, 94),
1689         c(29, 32, 63, 0, 93, 8, 71, 56, 8, 33, 84, 93, 69),
1690         c(70, 72, 64, 93, 0, 101, 56, 56, 92, 81, 16, 5, 69),
1691         c(35, 40, 68, 8, 101, 0, 76, 62, 11, 39, 91, 101, 76),
1692         c(42, 71, 11, 71, 56, 76, 0, 15, 65, 81, 40, 60, 94),
1693         c(27, 56, 12, 56, 56, 62, 15, 0, 50, 66, 41, 58, 82),
1694         c(24, 36, 56, 8, 92, 11, 65, 50, 0, 39, 81, 91, 74),
1695         c(44, 11, 77, 33, 81, 39, 81, 66, 39, 0, 77, 79, 37),
1696         c(58, 66, 48, 84, 16, 91, 40, 41, 81, 77, 0, 20, 73),
1697         c(71, 70, 67, 93, 5, 101, 60, 58, 91, 79, 20, 0, 65),
1698         c(69, 37, 94, 69, 69, 76, 94, 82, 74, 37, 73, 65, 0)
1699  )
1700  cost <- get_cost(
1701    tour = initial_candidate,
1702    distance_matrix = distance_matrix)
1703  print(cost)
```

R output

```
1704  [1] 709
```

Now a function for a neighbourhood operator will be written, swap_stops: swapping two stops in a given tour.

```
R input
1705   #' Return a new tour by swapping two stops.
1706   #'
1707   #' @param tour A given vector of successive stops.
1708   #'
1709   #' @return A tour
1710   swap_stops <- function(tour){
1711     number_of_stops <- length(tour) - 1
1712     stops_to_swap <- sample(2:number_of_stops, 2)
1713     new_tour <- replace(
1714       x = tour,
1715       list = stops_to_swap,
1716       values = rev(tour[stops_to_swap])
1717     )
1718   }
```

Applying this neighbourhood operator to the initial candidate gives:

```
R input
1719   new_tour <- swap_stops(initial_candidate)
1720   print(new_tour)
```

which swaps the 6th and 11th stops:

```
R output
1721   [1]  0  9  4  7  1 11  5  3  8  6  2 12 10  0
```

Now all the tools are in place to build a tool to carry out the neighbourhood search `run_neighbourhood_search`.

```
R input
1722   #' Returns a tour by carrying out a neighbourhood search
1723   #'
1724   #' @param distance_matrix: the distance matrix
1725   #' @param iterations: the number of iterations for
1726   #'                    which to run the algorithm
1727   #' @param seed: a random seed (default: None)
1728   #' @param neighbourhood_operator: the neighbourhood operation
1729   #'                               (default: swap_stops)
```

```
1730   #'
1731   #' @return A tour
1732   run_neighbourhood_search <- function(
1733     distance_matrix,
1734     iterations,
1735     seed = NA,
1736     neighbourhood_operator = swap_stops
1737   ){
1738     number_of_stops <- nrow(distance_matrix)
1739     candidate <- get_initial_candidate(
1740       number_of_stops = number_of_stops,
1741       seed = seed
1742     )
1743     best_cost <- get_cost(
1744       tour = candidate,
1745       distance_matrix = distance_matrix
1746     )
1747     for (repetition in 1:iterations) {
1748       new_candidate <- neighbourhood_operator(candidate)
1749       cost <- get_cost(
1750         tour = new_candidate,
1751         distance_matrix = distance_matrix
1752       )
1753       if (cost <= best_cost) {
1754         best_cost <- cost
1755         candidate <- new_candidate
1756       }
1757     }
1758     candidate
1759   }
```

Now running this for 1000 iterations:

R input

```
1760   number_of_iterations <- 1000
1761   solution_with_swap_stops <- run_neighbourhood_search(
1762     distance_matrix = distance_matrix,
1763     iterations = number_of_iterations,
1764     seed = seed,
1765     neighbourhood_operator = swap_stops
```

```
1766  )
1767  print(solution_with_swap_stops)
```

gives:

```
R output
1768  [1]  0 11  4 10  6  2  7 12  9  1  3  5  8  0
```

This has a cost:

```
R input
1769  cost <- get_cost(
1770    tour = solution_with_swap_stops,
1771    distance_matrix = distance_matrix
1772  )
1773  print(cost)
```

which gives:

```
R output
1774  [1] 360
```

Therefore, using this particular algorithm, a pretty good route is found, with a total distance of 360.

It is important to note that this may not be the optimal route, and different algorithms may produce better solutions. For example, one way to modify the algorithm is to use a different neighbourhood operator. Instead of swapping two stops, reverse the path between those two stops. This corresponds to an algorithm called the "2-opt" algorithm[3]. The reverse_path function does this:

```
R input
1775  #' Return a new tour by reversing the path between two stops.
1776  #'
1777  #' @param tour A given vector of successive stops.
1778  #'
1779  #' @return A tour
1780  reverse_path <- function(tour){
```

---

[3]The 2 opt algorithm was first published in [13].

```
1781    number_of_stops <- length(tour) - 1
1782    stops_to_swap <- sample(2:number_of_stops, 2)
1783    i <- min(stops_to_swap)
1784    j <- max(stops_to_swap)
1785    new_order <- c(c(1: (i - 1)), c(j:i), c( (j + 1): length(tour)))
1786    tour[new_order]
1787  }
```

Applying this neighbourhood operator to the initial candidate gives:

R input

```
1788  new_tour <- reverse_path(initial_candidate)
1789  print(new_tour)
```

which reverses the order between the 3rd and the 13th stop:

R output

```
1790    [1]  0  9 10 12 11  6  8  3  5  2  1  7  4  0
```

Now running the neighbourhood search for 1000 iterations using the reverse_path neighbourhood operator:

R input

```
1791  number_of_iterations <- 1000
1792  solution_with_reverse_path <- run_neighbourhood_search(
1793    distance_matrix = distance_matrix,
1794    iterations = number_of_iterations,
1795    seed = seed,
1796    neighbourhood_operator = reverse_path
1797  )
1798  print(solution_with_reverse_path)
```

gives:

R output

```
1799    [1]  0  7  2  6 10  4 11 12  9  1  3  5  8  0
```

This now gives a different route. Importantly, the costs differ substantially:

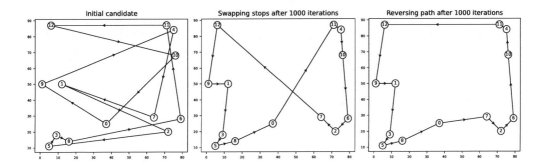

Figure 9.5 The final tours obtained by using the neighbourhood search in R.

```
1800  cost <- get_cost(
1801      tour = solution_with_reverse_path,
1802      distance_matrix = distance_matrix
1803  )
1804  print(cost)
```

which gives:

```
1805  [1] 299
```

This is an improvement on the solution found using the `swap_stops` operator. Figure 9.5 shows the final obtained routes given by both approaches.

9.5 WIDER CONTEXT

Heuristic methods, sometimes referred to as meta-heuristics, are a whole family of algorithms used to find approximate solutions to combinatorial optimisation problems. An overview is given in [4]. These algorithms include greedy searches, tabu searches, simulated annealing, genetic algorithms as well as ant colony optimisation. They are usually employed when the problem is too large or complex to use exact methodologies.

The travelling salesman problem, described in this chapter, is a classic example of one of these problems, formally described first in [41], although thought to have been discussed informally centuries before. It is an example of a large number of types of problems collectively known as vehicle routing problems, which often require heuristic methods for their solutions. A survey is given in [5]. Variations of the problem include multiple, heterogeneous and/or capacitated vehicles, and stochastic or time-dependent travel times. A recent adaptation of the problem is the green vehicle routing problem [45], where the cost function includes consideration of greenhouse gas emissions and other pollutants.

For more diverse applications of heuristic methods, consider [36], which describes a tabu search algorithm for finding seating plans for a wedding; and [75] where a genetic algorithm is used to build a prediction model for locations of deep-sea wildlife habitats.

Bibliography

[1] Robert Axelrod. *The Evolution of Co-Operation*. Penguin Books, 1990. ISBN: 9780140124958.

[2] Stefan Milton Bache and Hadley Wickham. *magrittr: A Forward-Pipe Operator for R*. R package version 2.0.1. 2020. URL: https://CRAN.R-project.org/package=magrittr.

[3] Hans W. Borchers. *pracma: Practical Numerical Math Functions*. R package version 2.3.3. 2021. URL: https://CRAN.R-project.org/package=pracma.

[4] Omid Bozorg-Haddad, Mohammad Solgi, and Hugo A. Loáiciga. *Meta-heuristic and evolutionary algorithms for engineering optimization*. John Wiley & Sons, 2017.

[5] Kris Braekers, Katrien Ramaekers, and Inneke Van Nieuwenhuyse. "The vehicle routing problem: State of the art classification and review". In: *Computers & Industrial Engineering* 99 (2016), pp. 300–313.

[6] Sally C. Brailsford et al. "An analysis of the academic literature on simulation and modelling in health care". In: *Journal of Simulation* 3.3 (2009), pp. 130–140.

[7] Sally C. Brailsford et al. "Hybrid simulation modelling in operational research: A state-of-the-art review". In: *European Journal of Operational Research* 278.3 (2019), pp. 721–737. ISSN: 0377-2217. DOI: https://doi.org/10.1016/j.ejor.2018.10.025. URL: https://www.sciencedirect.com/science/article/pii/S0377221718308786.

[8] Richard L. Burden, J. Douglas Faires, and Albert C. Reynolds. *Numerical analysis*. Brooks/cole Pacific Grove, CA, 2001.

[9] Pedro Cavalcante Oliveira, Diego S. Cardoso, and Marcelo Gelati. *Recon: Computational Tools for Economics*. R package version 0.3.0.0. 2019. URL: https://CRAN.R-project.org/package=Recon.

[10] Michele Conforti, Gérard Cornuéjols, Giacomo Zambelli, et al. *Integer programming*. Vol. 271. Springer, 2014.

[11] Ian Cooper, Argha Mondal, and Chris G. Antonopoulos. "A SIR model assumption for the spread of COVID-19 in different communities". In: *Chaos, Solitons & Fractals* 139 (2020), p. 110057.

[12] J.M. Coyle, D. Exelby, and J. Holt. "System dynamics in defence analysis: some case studies". In: *Journal of the Operational Research Society* 50.4 (1999), pp. 372–382.

[13] Georges A. Croes. "A method for solving traveling-salesman problems". In: *Operations Research* 6.6 (1958), pp. 791–812.

[14] Sarang Deo and Itai Gurvich. "Centralized vs. decentralized ambulance diversion: A network perspective". In: *Management Science* 57.7 (2011), pp. 1300–1319.

[15] Diego A. Díaz, Ana María Jiménez, and Cristián Larroulet. "An agent-based model of school choice with information asymmetries". In: *Journal of Simulation* 15.1-2 (2021), pp. 130–147.

[16] Ocotlán Díaz-Parra et al. "A survey of transportation problems". In: *Journal of Applied Mathematics* 2014 (2014).

[17] Guillermo Durán et al. "Scheduling the Chilean soccer league by integer programming". In: *Interfaces* 37.6 (2007), pp. 539–552.

[18] Gonçalo Figueira and Bernardo Almada-Lobo. "Hybrid simulation–optimization methods: A taxonomy and discussion". In: *Simulation Modelling Practice and Theory* 46 (2014), pp. 118–134.

[19] Michael J. Flynn. "Very high-speed computing systems". In: *Proceedings of the IEEE* 54.12 (1966), pp. 1901–1909.

[20] John Forrest and Robin Lougee-Heimer. "CBC user guide". In: *Emerging theory, methods, and applications*. INFORMS, 2005, pp. 257–277.

[21] Jay W. Forrester. "Industrial dynamics." In: *Pegasus Communications, Waltham, MA* (1961).

[22] Jay W. Forrester. "The beginning of system dynamics". In: *McKinsey Quarterly* (1995).

[23] Drew Fudenberg et al. *The theory of learning in games*. Vol. 2. MIT press, 1998.

[24] Martin Gardener. "MATHEMATICAL GAMES: The fantastic combinations of John Conway's new solitaire game life". In: *Scientific American* 223.4 (1970), pp. 120–123.

[25] Vincent Goulet et al. *expm: Matrix Exponential, Log, 'etc'*. R package version 0.999-6. 2021. URL: https://CRAN.R-project.org/package=expm.

[26] Jeff D. Griffiths, Janet E. Williams, and Richard Max Wood. "Modelling activities at a neurological rehabilitation unit". In: *European Journal of Operational Research* 226.2 (2013), pp. 301–312.

[27] Francesca Guerriero and Rosita Guido. "Operational research in the management of the operating theatre: a survey". In: *Health Care Management Science* 14.1 (2011), pp. 89–114.

[28] Charles R. Harris et al. "Array programming with NumPy". In: *Nature* 585.7825 (Sept. 2020), pp. 357–362. DOI: 10.1038/s41586-020-2649-2. URL: https://doi.org/10.1038/s41586-020-2649-2.

[29] Mhand Hifi et al. "A linear programming approach for the three-dimensional bin-packing problem". In: *Electronic Notes in Discrete Mathematics* 36 (2010), pp. 993–1000.

[30] J. D. Hunter. "Matplotlib: A 2D graphics environment". In: *Computing in Science & Engineering* 9.3 (2007), pp. 90–95. DOI: `10.1109/MCSE.2007.55`.

[31] Peng Jing et al. "Agent-based simulation of autonomous vehicles: A systematic literature review". In: *IEEE Access* 8 (2020), pp. 79089–79103.

[32] Auguste Kerckhoffs. "La cryptographie militaire. Journal des sciences militaries". In: *IX (38)* 5 (1883).

[33] Thomas Kluyver et al. "Jupyter Notebooks – a publishing format for reproducible computational workflows". In: *Positioning and Power in Academic Publishing: Players, Agents and Agendas.* Ed. by F. Loizides and B. Schmidt. IOS Press. 2016, pp. 87–90.

[34] Vincent Knight and James Campbell. "Nashpy: A Python library for the computation of Nash equilibria". In: *Journal of Open Source Software* 3.30 (2018), p. 904. DOI: `10.21105/joss.00904`. URL: `https://doi.org/10.21105/joss.00904`.

[35] Frederick William Lanchester. *Aircraft in warfare: The dawn of the fourth arm.* Constable limited, 1916.

[36] Rhyd Lewis and Fiona Carroll. "Creating seating plans: a practical application". In: *Journal of the Operational Research Society* 67.11 (2016), pp. 1353–1362.

[37] Juan Martín García. *Theory and practical exercises of System Dynamics.* 2018.

[38] Michael Maschler, Eilon Solan, and Shmuel Zamir. *Game theory.* Vol. 979. 2013, p. 4.

[39] Makoto Matsumoto and Takuji Nishimura. "Mersenne twister: a 623-dimensionally equidistributed uniform pseudo-random number generator". In: *ACM Transactions on Modeling and Computer Simulation (TOMACS)* 8.1 (1998), pp. 3–30.

[40] Wes McKinney. "Data Structures for Statistical Computing in Python". In: *Proceedings of the 9th Python in Science Conference.* Ed. by Stéfan van der Walt and Jarrod Millman. 2010, pp. 56–61. DOI: `10.25080/Majora-92bf1922-00a`.

[41] Karl Menger. "Das botenproblem". In: *Ergebnisse eines Mathematischen Kolloquiums* 2 (1932), pp. 11–12.

[42] Aaron Meurer et al. "SymPy: symbolic computing in Python". In: *PeerJ Computer Science* 3 (Jan. 2017), e103. ISSN: 2376-5992. DOI: `10.7717/peerj-cs.103`. URL: `https://doi.org/10.7717/peerj-cs.103`.

[43] Zbigniew Michalewicz and David B Fogel. *How to solve it: modern heuristics.* Springer Science & Business Media, 2013.

[44] Stuart Mitchell, Michael OSullivan, and Iain Dunning. "PuLP: a linear programming toolkit for python". In: *The University of Auckland, Auckland, New Zealand* (2011), p. 65.

[45] Reza Moghdani et al. "The green vehicle routing problem: A systematic literature review". In: *Journal of Cleaner Production* 279 (2021), p. 123691.

[46] Cleve Moler and Charles Van Loan. "Nineteen dubious ways to compute the exponential of a matrix". In: *SIAM Review* 20.4 (1978), pp. 801–836.

[47] Cleve Moler and Charles Van Loan. "Nineteen dubious ways to compute the exponential of a matrix, twenty-five years later". In: *SIAM Review* 45.1 (2003), pp. 3–49.

[48] Sylvia Nasar. *A beautiful mind.* Simon and Schuster, 2011.

[49] John Nash. "Non-cooperative games". In: *Annals of Mathematics* (1951), pp. 286–295.

[50] John F. Nash et al. "Equilibrium points in n-person games". In: *Proceedings of the National Academy of Sciences* 36.1 (1950), pp. 48–49.

[51] Andres F. Osorio et al. "Simulation-optimization model for production planning in the blood supply chain". In: *Health Care Management Science* 20.4 (2017), pp. 548–564.

[52] Geraint I. Palmer et al. "Ciw: An open-source discrete event simulation library". In: *Journal of Simulation* 13.1 (2019), pp. 68–82. DOI: `10.1080/17477778.2018.1473909`.

[53] William H. Press and Freeman J. Dyson. "Iterated Prisoner's Dilemma contains strategies that dominate any evolutionary opponent". In: *Proceedings of the National Academy of Sciences* 109.26 (2012), pp. 10409–10413.

[54] Craig W. Reynolds. "Flocks, herds and schools: A distributed behavioral model". In: *Proceedings of the 14th annual conference on Computer graphics and interactive techniques.* 1987, pp. 25–34.

[55] Clark Robinson. *Dynamical systems: stability, symbolic dynamics, and chaos.* CRC press, 1998.

[56] Stewart Robinson. *Simulation: the practice of model development and use.* Vol. 50. Wiley Chichester, 2004.

[57] Iza Romanowska et al. "Agent-based modeling for archaeologists: Part 1 of 3". In: *Advances in Archaeological Practice* 7.2 (2019), pp. 178–184.

[58] Simo Särkkä and Arno Solin. *Applied stochastic differential equations.* Vol. 10. Cambridge University Press, 2019.

[59] Thomas C. Schelling. *Micromotives and macrobehavior.* WW Norton & Company, 2006.

[60] J.A.M. Schreuder. "Application of a location model to fire stations in Rotterdam". In: *European Journal of Operational Research* 6.2 (1981), pp. 212–219.

[61] Dirk Schumacher and Jeroen Ooms. *rcbc: COIN CBC MILP Solver Bindings.* R package version 0.1.0.9001. 2021.

[62] Yoav Shoham and Kevin Leyton-Brown. *Multiagent systems: Algorithmic, game-theoretic, and logical foundations.* Cambridge University Press, 2008.

[63] J. Maynard Smith. "The theory of games and the evolution of animal conflicts". In: *Journal of Theoretical Biology* 47.1 (1974), pp. 209–221.

[64] Karline Soetaert, Thomas Petzoldt, and R. Woodrow Setzer. "Solving Differential Equations in R: Package deSolve". In: *Journal of Statistical Software* 33.9 (2010), pp. 1–25. DOI: `10.18637/jss.v033.i09`.

[65] Ian Stewart. "Monopoly revisited". In: *Scientific American* 275.4 (1996), pp. 116–119.

[66] James Stewart. *Calculus: Concepts and contexts*. Cengage Learning, 2009.

[67] William J. Stewart. *Probability, Markov chains, queues, and simulation*. Princeton university press, 2009.

[68] Alan Sultan. *Linear programming: An introduction with applications*. Elsevier, 2014.

[69] Richard Syms and Laszlo Solymar. "A dynamic competition model of regime change". In: *Journal of the Operational Research Society* 66.11 (2015), pp. 1939–1947.

[70] Bari Tan. "Markov chains and the RISK board game". In: *Mathematics Magazine* 70.5 (1997), pp. 349–357.

[71] The pandas development team. *pandas-dev/pandas: Pandas*. Version latest. Feb. 2020. DOI: `10.5281/zenodo.3509134`. URL: `https://doi.org/10.5281/zenodo.3509134`.

[72] The Ciw library developers. *Ciw: 2.2.0*. 2021. URL: `http://dx.doi.org/10.5281/zenodo`.

[73] The Simpy developers. *Simpy documentation*. Oct. 2021. URL: `https://simpy.readthedocs.io/en/latest/`.

[74] Stefan TheuSSl, Florian Schwendinger, and Kurt Hornik. *ROI: The R Optimization Infrastructure Package*. Research Report Series / Department of Statistics and Mathematics 133. Vienna: WU Vienna University of Economics and Business, Oct. 2017. URL: `http://epub.wu.ac.at/5858/`.

[75] Ruiju Tong et al. "Modeling the habitat suitability for deep-water gorgonian corals based on terrain variables". In: *Ecological Informatics* 13 (2013), pp. 123–132.

[76] Iñaki Ucar, Bart Smeets, and Arturo Azcorra. "simmer: Discrete-Event Simulation for R". In: *Journal of Statistical Software* 90.2 (2019), pp. 1–30. DOI: `10.18637/jss.v090.i02`.

[77] Charles F. Van Loan and G. Golub. *Matrix computations (Johns Hopkins studies in mathematical sciences)*. The Johns Hopkins University Press, 1996.

[78] James S. Vandergraft. "A fluid flow model of networks of queues". In: *Management Science* 29.10 (1983), pp. 1198–1208.

[79] Jesús Isaac Vázquez-Serrano and R.E. Peimbert-Garca. "System dynamics applications in healthcare: A literature review". In: *Proceedings of the international conference on industrial engineering and operations management*. 2020, pp. 10–12.

[80] Pauli Virtanen et al. "SciPy 1.0: Fundamental Algorithms for Scientific Computing in Python". In: *Nature Methods* 17 (2020), pp. 261–272. DOI: `10.1038/s41592-019-0686-2`.

[81] John Von Neumann. "13. various techniques used in connection with random digits". In: *Applied Mathematics Series* 12.36-38 (1951), p. 3.

[82] Douglas J. White. "A survey of applications of Markov decision processes". In: *Journal of the Operational Research Society* 44.11 (1993), pp. 1073–1096.

[83] Hadley Wickham. *ggplot2: Elegant Graphics for Data Analysis*. Springer-Verlag New York, 2016. ISBN: 978-3-319-24277-4. URL: `https://ggplot2.tidyverse.org`.

[84] Greg Wilson et al. "Best practices for scientific computing". In: *PLoS Biology* 12.1 (2014), e1001745.

Index

Printed in the United States
by Baker & Taylor Publisher Services